# MY LIFE, MY SCIENCE

Pursuing a Cure for Huntington's Disease

OTHER TITLES FROM COLD SPRING HARBOR LABORATORY PRESS

*Conscience and Courage: How Visionary CEO Henri Termeer Built a Biotech Giant and Pioneered the Rare Disease Industry*

*Orphan: The Quest to Save Children with Rare Genetic Disorders*

# MY LIFE, MY SCIENCE

## Pursuing a Cure for Huntington's Disease

NANCY SABIN WEXLER

in collaboration with Mark Hampton
and Alice R. Wexler

CSH PRESS
COLD SPRING HARBOR LABORATORY PRESS
Cold Spring Harbor, New York • www.cshlpress.org

MY LIFE, MY SCIENCE
Pursuing a Cure for Huntington's Disease



| | |
|---|---|
| **Acquisition Editor and Publisher** | John Inglis |
| **Project Manager** | Barbara Acosta |
| **Editorial Assistant** | Danett Gil |
| **Permissions Coordinator** | Carol Brown |
| **Production Editor** | Kathleen Bubbeo |
| **Production Manager and Cover Designer** | Denise Weiss |

*Front cover image:* Pedigree with Polaroids of Venezuelan families with Huntington's, September 1982. © Steve Uzzell, All Rights Reserved.

ISBN 978-1-621825-45-6 (hardcover)
ISBN 978-1-621825-46-3 (epub)

# CONTENTS

# PREFACE

Nancy began writing this memoir in early 2020, soon after she was diagnosed with Huntington's disease. A month later, the world was diagnosed with COVID. Although we were already collaborating on her project, our work became more difficult. As New York City shut down, Nancy, at age 75 and living with HD, learned to use and love Zoom. When the mRNA vaccines proved effective and shutdowns lifted, more than a year had gone by. Some months later, Nancy decided to invite her sister, Alice, an experienced writer, to join in helping her tell her story. The three of us then worked together, partly in person, partly by Zoom, and partly by email and phone. With Nancy's speech somewhat compromised, although not her memory nor her intellect, it fell to Alice and me to put her words on the page. But dictating memories, providing correspondence and interview transcripts, correcting errors, and revising interpretations, Nancy has remained throughout the teller of her own story, the author of her own extraordinary life.

**Mark Hampton**

# PROLOGUE

*July 1979, State of Zulia, western Venezuela*

As we glide through the waters of Lake Maracaibo, daubs of color appear in the distance like a hallucination or a Matisse painting. From our *chalana*—a large canoe fitted with an outboard motor—we make out a scattering of small houses splashed with paint and perched on stilts over the water. With my colleague Tom Chase, a neurologist at the National Institutes of Health where I work as a health sciences administrator, I have come on an exploratory mission to the remote village of Laguneta. We are searching for someone who has inherited Huntington's disease from both of their parents, called a homozygote. Known here as *el mal de San Vito* or simply *el mal,* Huntington's disease (HD) causes involuntary jerky movements of the body known as chorea, mental disturbances, and cognitive decline, leading inexorably to death over ten or twenty or more years. Just one parent with the illness can transmit it to their children, who each have a 50–50 chance of inheriting the disease.

We are hoping that an individual with a double inheritance of this malady will reveal its fundamental cause, as happened a few years earlier with a hereditary condition known as hypercholesterolemia, which causes early heart attacks and strokes.[1] We figure our best chance of finding such a "double dose" individual is to look for families in which both parents have *el mal* and then check out their children for atypically severe or unusual disease. But we have been in Venezuela for nearly a week and have not yet found such families, much less anyone whose Huntington's stands out as unique.

As we round a spit of land and enter a lagoon, the *pueblo de agua,* or village over water, comes into view. Laguneta is comprised of twenty-five

*Approaching Laguneta. (Photo © Steve Uzzell, All Rights Reserved.)*

or so small houses built on stilts in the lake, most of them constructed of wood or tin covered by corrugated tin roofs, with occasional windows framed in bright green or blue. Each house boasts a wooden porch, many adorned with piles of fishing nets and mounds of dried salted fish, testimony to the major occupation in this community. Here on this porch is a pig in a pen; over there, a pet monkey; farther on, a cormorant tied by the leg. Red, green, yellow, and orange hammocks hang like butterflies, cradling infants and adults, while naked children with bellies protruding from malnutrition and parasites, heads shaved clean to keep cool and avoid lice, cling shyly to the sides of the houses or plunge into the water, twisting like tops and shrieking with joy.

We have come to Laguneta owing to the work of a Venezuelan physician, Dr. Américo Negrette, whose pioneering studies, in the 1950s and 1960s, of Huntington's disease in Venezuela laid the foundation for all that followed. Now, decades later, he welcomed us on our arrival in Maracaibo, walking us through the barrio of San Luis, on the outskirts of the city, where he once cared for many families with *el mal*. Because they love him, they open their homes to us, in San Luis and in the villages of Barranquitas and Laguneta, the three major centers of the disease in this region. Nearly half the people in Laguneta suffer from *el mal* or are at 50% risk, probably owing to little migration out due to poverty and

*Family dwelling in Laguneta. (Photo © Steve Uzzell, All Rights Reserved.)*

isolation, and intermarriage among the affected families, although no one knows for certain.

Three things the people here do know: That *el mal* runs in families (is hereditary), and both women and men can pass on the disease to their children, although not all children of a parent with the disease will inevitably get it. Yet they know that just one parent with *el mal* means that one or more of the kids will likely develop it too. Like my grandfather, uncles, and most recently my mother Leonore, who began showing symptoms decades ago and has been gone from this earth for a little more than a year.

As the clock ticks toward our departure, we intensify our search, hoping to find our homozygote before we must leave for home, possibly never to return.

♦ ♦ ♦

Our guide cuts the motor, and we glide up to one house in the center of the village. People freeze and gape at us in silence. Tom, six-feet-two, thin, and pale, and I—*la catira* they would call me, the blonde—with my long hair and blue eyes, arrive in this place like visitors from another galaxy. A woman catches my eye as she sits cross-legged on the porch. She had been cleaning fish, but seeing us, she stops. She wears a gray cotton dress that hangs straight down from her shoulders. Her long black hair is pulled back into a ponytail. She has brown eyes, a strong jaw, and when she finally smiles, a few scraggly teeth. At first, she remains totally still. Then her strained effort gives way to a series of jerky movements. I know this woman; I know how she will get up with a lurch, walk with a wide gait to resist her movements, slur her words that I might not understand but I will know what she is saying anyway. She will move her mouth and face and eyes against her will so that they speak for her. She will even feel the same as my mother to hug, with her muscles constantly working to escape my grasp, even as she won't let go.

Our Lady of Laguneta holds me in her spell. The continuous movement accompanying her words has a hypnotic effect as she greets us. We describe our project, and she tells us about her thirteen children and her husband who had been at risk but has recently died of pneumonia, leaving her to raise her brood alone. Crestfallen, we explain that for our research, we need to find someone with two *living* parents with *el mal*. As we thank her and start to leave, she tells us, almost as an afterthought, "I have a sister...". That sister lives with her husband and children in a house a little farther along the grove of houses in the lagoon. She gestures toward them. "Go talk to them, they'll help you," she says.

At her sister's house, there are more hammocks, more children scurrying about. We climb out of our *chalana* onto the porch. Inside the house we meet the sister, a chain-smoking woman with sharp features, elegant bearing, and obvious chorea. She eyes us with a mixture of curiosity and suspicion. We introduce ourselves by way of her sister and tell her why we have come. The woman smiles and her reserve dissolves in the warmth of that smile. Then we see a white T-shirt moving in a hammock strung up at the other end of the room; clearly the man inside that T-shirt has Huntington's as well. He gets up out of the hammock to greet us. Soon he, too, is in the familiar state of standing still and moving at the same time. At first, we think they are brother and sister, but no, they are married, with fourteen children, at least one of whom has likely inherited the faulty gene from both parents.

We have finally found the family we were looking for. I should have felt elated. Instead, I am overcome by feelings of sadness, as I realize that most of these beautiful children will develop the disease, inheriting either one or two copies of the faulty gene. Satisfaction and sorrow are intertwined within me like the two strands of the double helix. My thoughts of Venezuela will be forever etched with memories of these bright, vivacious kids, many of whose lives will fade and be cut short due to the cruelty of a single dominant gene.

♦ ♦ ♦

The children, timid at first, grow braver and give in to their curiosity as they crowd around us. The woman's smile fades slightly as she puts her hands on the shoulders of a teenage girl, her daughter. "One is already sick," she says. We talk to the children, and they agree, albeit with trepidation, to let Tom draw their blood while I caress them and speak to them in my pidgin Spanish. After a while, a man poles up to the house in his *chalana*. He is the chain-smoker's brother and the chieftain of the town, we are told. As I discover later, he is also one of the most intelligent men I've ever met. He appears to be in the very early stage of *el mal*, with slight jerky movements and a hint of unsteadiness but no other signs of anything amiss. He wants us to meet his son. He poles away and brings back an angelic boy of five or six, with enormous blue-green eyes, curly light brown hair, freckles, and a slow, incandescent smile like a Cheshire cat's. The chain-smoker introduces her nephew. I think he has Huntington's, Tom says to me in English. The child has been ill since age two and now moves as if through glue. We watch him carefully as he walks like a little old man, falling, taking another step, falling again. He is certainly the youngest case of Huntington's that we have encountered so far.

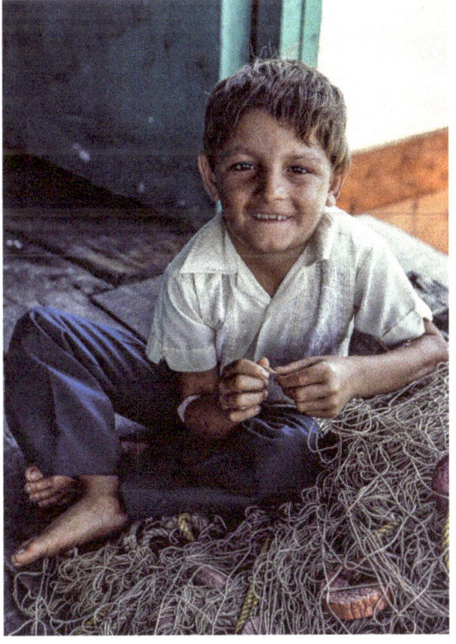

*Always in my heart. (Photo © Steve Uzzell, All Rights Reserved.)*

His aunt tells us who does and does not have *el mal* in their family, weaving together a complex family history going back multiple generations, a history that will eventually become crucial to our project. She and her husband also agree to donate blood. That night we return to Maracaibo and the next day, we will leave for home, carrying with us, along with precious blood samples, bittersweet memories of our first visit to this haunted place and the children whom we cannot forget. Especially that little boy who will die in the throes of uncontrollable seizures at the age of 11, but not before making his contribution to the world in a way that will surprise us all.

CHAPTER

# 1

# FAMILY SECRETS

*Summertime and the livin' is easy. Fish are jumpin' and the cotton is high.*

Ira Gershwin

It's July 1959 and I'm waterskiing in my bikini behind our Chris-Craft on Lake Tahoe, high in the Sierras of Northern California, singing at the top of my lungs, "Oh, your daddy's rich and your ma is good lookin'...". That song could almost have been written about fourteen-year-old me. My father's office was in Beverly Hills, on North Roxbury Drive, and Gershwin lived just down the street. Dad could have known him. Milton Wexler was becoming a psychoanalyst to the stars, and in 1959 Beverly Hills was still a small town.

We were living then in Pacific Palisades and for a few years in the 1950s spent our summers at Lake Tahoe. Our wood-frame house at the edge of Carnelian Bay faced a small pier and boathouse below. We were all happier when we were there. Because my mother was happier. Or more accurately, she was less unhappy. In those golden summers, the wall of depression that would eventually seal her off from us was still penetrable. When friends came to visit, she could still sparkle and be gay. On any given night at Tahoe, or occasionally back home in Los Angeles, our dinner guests might include Andrés Segovia's protégé Michael "Tiger" Lorimer—my high school friend—playing classical guitar; my father's best friend and analyst colleague Ralph "Romi" Greenson and his wife Hildi; occasionally a young man struggling with schizophrenia; and when Romi was out of town, one of his patients: Marilyn Monroe, struggling with being Hollywood bombshell Marilyn Monroe. But whoever the guests happened to be, my most treasured memories of that time are

(Left to right) *Me, Dad, Alice, young friend, and Mom on the deck of our house at Lake Tahoe, ca. 1959.*

the occasions when my mother Leonore came out of her sadness for the evening, nights when our family was at its best.

Inevitably, the summer would come to an end, and we'd return to our home on Napoli Drive overlooking the Pacific Ocean in the distance, with Catalina Island just visible on the horizon. Dad would return to his busy practice, my sister Alice and I to school, and my mother to her depression—drinking coffee, smoking cigarettes, and staring out at paradise.

"Mom, can you help me with my homework?"

"Ask your father."

By the time I was in high school, the pain of engaging was palpable. Pressing my mother into conversation would have been cruel. I would simply hug her and hold her.

Even as a toddler, I had become accustomed to the micro-moments when Mom would seem to go elsewhere. A flash of blankness in her eyes. A lost beat in the conversation. And then, almost before it happened, it passed and she returned. I felt keenly every one of those departures, even if I didn't know where she had gone or what had called her away. No one did then. Not even Dad.

She had never told him about visiting her father Abraham, confined to a state mental hospital on Long Island, suffering the end stage of Huntington's chorea—as they called it then—the last image of him she had carried with her since age fifteen, in 1929, when he died. This undercurrent of sadness she kept to herself, saying nothing even when she must have seen her three older brothers begin to show the telltale jerky

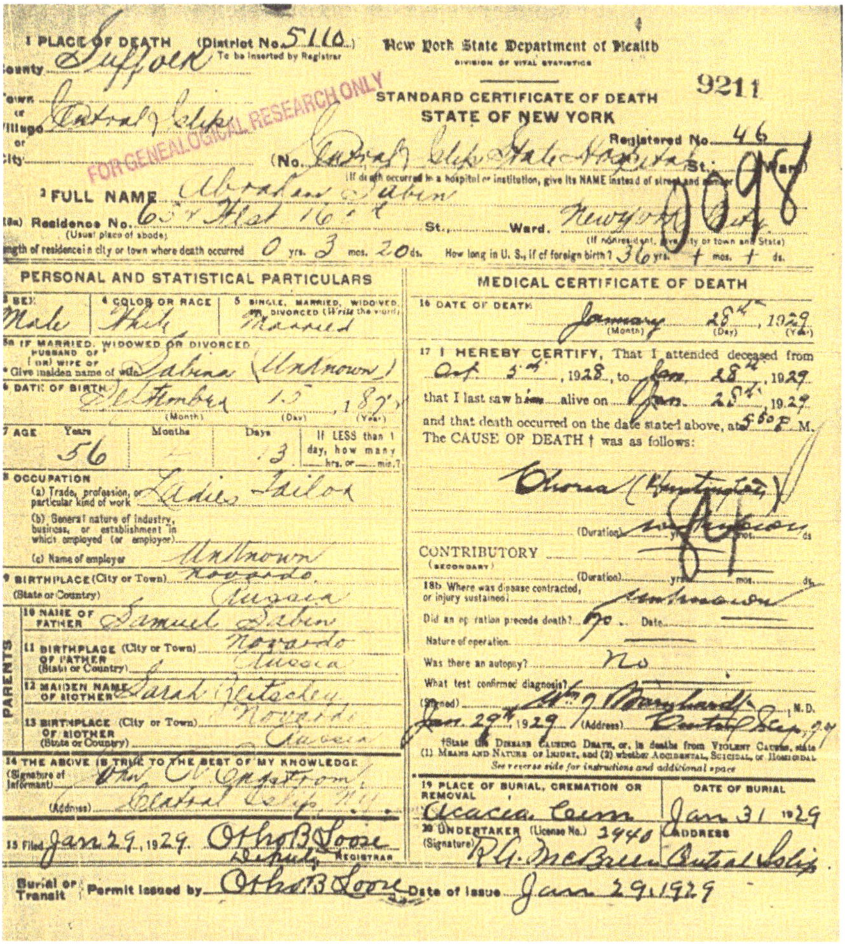
New York State Department of Health
DIVISION OF VITAL STATISTICS
STANDARD CERTIFICATE OF DEATH
STATE OF NEW YORK
9211
FOR GENEALOGICAL RESEARCH ONLY

1 PLACE OF DEATH (District No. 5110) To be inserted by Registrar
County Suffolk
Town or Village or City Central Islip
Registered No. 46
(No. Central Islip State Hospital St., Ward)
(If death occurred in a hospital or institution, give its NAME instead of street and number)
2 FULL NAME Abraham Sabin
(2a) Residence No. 65 West 16th St., Ward. New York City
(Usual place of abode) (If nonresident, give city or town and State)
Length of residence in city or town where death occurred 0 yrs. 3 mos. 20 ds. How long in U. S., if of foreign birth? 36 yrs. mos. ds.

PERSONAL AND STATISTICAL PARTICULARS
3 SEX Male
4 COLOR OR RACE White
5 SINGLE, MARRIED, WIDOWED, OR DIVORCED (Write the word) Married
5a IF MARRIED, WIDOWED OR DIVORCED HUSBAND OF (OR) WIFE OF *Give maiden name of wife Sabina (Unknown)
6 DATE OF BIRTH September 15, 1872 (Month) (Day) (Year)
7 AGE Years 56 Months Days 13 If LESS than 1 day, how many hrs. or min.?
8 OCCUPATION
(a) Trade, profession, or particular kind of work Ladies Tailor
(b) General nature of industry, business, or establishment in which employed (or employer)
(c) Name of employer Unknown
9 BIRTHPLACE (City or Town) Novardo
(State or Country) Russia
PARENTS
10 NAME OF FATHER Samuel Sabin
11 BIRTHPLACE OF FATHER (City or Town) Novardo (State or Country) Russia
12 MAIDEN NAME OF MOTHER Sarah
13 BIRTHPLACE OF MOTHER (City or Town) Novardo (State or Country) Russia
14 THE ABOVE IS TRUE TO THE BEST OF MY KNOWLEDGE
(Signature of Informant)
(Address) Central Islip, N.Y.
15 Filed Jan 29, 1929. Otho B. Loose Deputy Registrar

MEDICAL CERTIFICATE OF DEATH
16 DATE OF DEATH January 28th, 1929 (Month) (Day) (Year)
17 I HEREBY CERTIFY, That I attended deceased from Oct. 5th, 1928, to Jan. 28th, 1929, that I last saw him alive on Jan. 28th, 1929, and that death occurred on the date stated above, at 5:50 P. M.
The CAUSE OF DEATH † was as follows:
Chorea (Huntington)
(Duration) unknown yrs. mos. ds.
CONTRIBUTORY (SECONDARY) (Duration) yrs. mos. ds.
18b Where was disease contracted, or injury sustained? Unknown
Did an operation precede death? No Date
Nature of operation
Was there an autopsy? No
What test confirmed diagnosis?
(Signed) , M.D.
Jan. 29th, 1929 (Address) Central Islip, N.Y.
†State the DISEASE CAUSING DEATH, or, in deaths from VIOLENT CAUSES, state (1) MEANS AND NATURE OF INJURY, and (2) whether ACCIDENTAL, SUICIDAL, or HOMICIDAL
See reverse side for instructions and additional space
19 PLACE OF BURIAL, CREMATION OR REMOVAL Acacia Cem. DATE OF BURIAL Jan 31 1929
20 UNDERTAKER (License No.) 3440 (Signature) R. G. McBreen ADDRESS Central Islip
Burial or Transit Permit issued by Otho B. Loose Date of issue Jan 29, 1929

*Death certificate of Abraham Sabin at Central Islip State Hospital, Long Island, New York, in 1929, showing cause of death as "Chorea (Huntington), duration unknown."*

movements that she'd seen her father exhibit years before. She would later explain that she'd felt no need to tell Dad about her father because she believed that Huntington's affected only men and therefore would not touch her or their children. She said she had overheard a doctor at the hospital tell her mother that Abraham had died of Huntington's chorea, so she went to the library and looked it up and read that it was a hereditary disease of men.

Mom's explanation, though, is something of an enigma. It is true that George Huntington himself, author of the classic 1872 description of the disease, had speculated that it affected slightly more men than women. And a few early twentieth-century neurologists agreed. Perhaps by the time she met Dad, in 1936, Mom's memory had translated "slightly more men than women" into her wished-for "only men." Yet the preponderance of medical and scientific writing about Huntington's chorea made no claims of gender preference. George Huntington had even described women with chorea out and about in his own hometown.[1]

Mom was fifteen when her father died, and his illness evidently inspired an interest in the new science of genetics and in biology more broadly. She majored in biology at Hunter College where she presided over the Biology Club, spending one summer on a scholarship at the Biological Station in Woods Hole, Massachusetts, and another at the University of Michigan Biological Station at Lake Cheboygan. She spent many hours at the American Museum of Natural History, a favorite haunt. After graduating with honors, she earned a Master's degree in Columbia University's illustrious zoology department, studying in the famous "fly room" established by Thomas Hunt Morgan, a founder of modern genetics. She might have gone on to earn a PhD like her friend Rose, except that my grandmother warned her she'd never get married if she did. She got a high school teaching credential instead.

Mom would have been familiar with Charles Davenport's widely used textbook, *Heredity in Relation to Eugenics,* which held up Huntington's as a prime example of a "hereditary defect" affecting both sexes. Here was a disease that the proponents of eugenics—a popular social and scientific movement for "better breeding"—cited as justification for sterilization. She would probably have learned the advice physicians gave to people—women and men alike—who had a parent with this disease: Don't get married, don't have children, and preferably get sterilized. How else could you eliminate the disease when abortion was illegal and birth control undependable and hard to access? Mom probably also knew the advice to anyone contemplating marrying into a family like hers: Don't.

It is not surprising, then, that most families with Huntington's tried to hide it. My grandfather Abraham had likely not told his wife Sabina, my grandmother, about his sister back in Russia who had developed the disease and committed suicide, as we found out much later. And why would he if he wanted to have a family? Sabina evidently never discussed it with Mom. So why would Mom tell her fiancé? In the early twentieth century, eugenics attracted the most up-to-date scientists and intellectuals as well as social leaders, even progressive ones. They believed they could "improve" the social fabric by excluding from the country certain classes of immigrants considered inferior and by legalizing sterilization without consent of those deemed "unfit," a broad category that included those with hereditary diseases in the family—diseases like Huntington's, families like ours. Mom had good reasons to hide Huntington's. I probably would have done the same.

*Mom and Dad in the late 1930s.*

♦ ♦ ♦

I was five years old in the winter of 1950 when the dam of secrecy broke. We were living in Topeka, Kansas where my father was on the staff of the Menninger Foundation, a widely respected psychiatric treatment and training center, the four of us—Mom, Dad, my older sister Alice, and me—all tucked away together in our wood-frame house with a white picket fence and a mulberry tree out front. Dad had found his passion in psychoanalysis, and my parents were very much in love. Our yard would freeze over in the winter, and my sister and I would watch with awe through the window as Mom and Dad skated together over the ice in perfect synchrony.

Before arriving in Topeka, Dad had practiced law in New York City, becoming disgusted with the corruption he encountered. He had long been interested in the new theories of psychoanalysis, and in 1938, thanks to Mom's income from teaching, he quit his law practice and enrolled in a PhD program in clinical psychology at Teachers College of Columbia University. He also completed a training analysis and took seminars with Theodor Reik, an associate of Freud's. He had not quite finished his PhD when we arrived in Topeka in 1946, but he immediately found himself in a place where he was treated with respect and surrounded by colleagues he revered. At Menninger's he saw hospitalized patients, taught residents in psychiatry, and did research on the treatment of schizophrenia with talk therapy (as opposed to the more invasive shock therapy and lobotomy),

*Our Topeka house on W. 20th Street, 1946–1951.*

*Colleagues and wives at Menninger's. (*Left to right*) Bottom row, Mom, David Rapaport, Martin Mayman (my future PhD mentor at Michigan), and Dad; top row, Maryline Barnard, Robert Holt, and the wives of Holt and Rapaport.*

an experience that left him with a lasting enthusiasm for research. He flourished within the local psychoanalytic community, an admired "lay" (nonmedical) analyst among the MDs. He also treated his first movie star. Besides training hundreds of young psychiatric residents, Menninger's was where Hollywood moguls sent troubled actors. One of them was Robert Walker, who became Dad's patient and later praised his treatment when he returned to Los Angeles, publicity that helped instantly fill Dad's roster of patients when we arrived a few years later.

You could have put our idyllic life into a snow globe. Then, one fateful week in November 1950, Mom's three brothers in New York City were all diagnosed with Huntington's. All three. In the same week. The diagnoses picked up the globe and turned it upside down, and when it had been turned upright and the snow had settled, everything inside had been dislodged.

♦ ♦ ♦

Mom was devastated by her brothers' diagnoses. She was close to all of them but especially to Seymour, the youngest, who played saxophone in the orchestra of his older brother Paul. She liked to tell us how they played at the big hotels in Miami and at Tavern on the Green, an iconic restaurant and music venue in Central Park.

*Paul Sabin*

*Seymour Sabin*

Jesse, the eldest and the only married brother, worked in retail. All three brothers had helped raise Mom when their father fell ill; they were proud and protective of their little sister, the only one of the four to graduate from college. The prospect of all three retracing their father's trajectory must have been almost too much for Mom to bear.

And there was more. It's unclear to me exactly how Mom and Dad both learned definitively that women as well as men were vulnerable to Huntington's. In his memoir, written more than forty years later, Dad stated that his brother Henry, an MD, told him in 1950, at the time of the Sabin brothers' diagnoses, that the disease impacted only men. He claimed that he didn't learn that Huntington's affects both sexes until Mom's diagnosis many years later (in 1968). At other times, though, Dad recalled that a surgeon neighbor in Topeka told him the truth when the Sabin brothers were diagnosed, a more convincing explanation. I was still a child, after all, and these explanations came to me much later, through the pipeline of family lore, a channel notoriously rife with unreliable witnesses.

Whatever the truth, it is difficult to fault my father for feeling betrayed by his wife's silence about her family legacy. The tapestry of trust had been ripped apart and would not be sewn together again. And yet, whether Mom knew the truth all along or learned it only later, it's also hard to fault her for wanting to keep Huntington's hidden, as had so many other families. And, for both of them, the knowledge that they had put their daughters' lives in jeopardy, and also the lives of any children we might have, opened a wound, never to be healed.

*Me in the mulberry tree.*

My father may not have believed in the old-fashioned "children should be seen and not heard," but he did believe parents should keep their troubles to themselves. He thought children should not be burdened with traumatic information unless absolutely necessary. Evidently Mom agreed. In their eyes, the diagnoses of Seymour, Paul, and Jesse Sabin did not rise to that level of urgency. Mom and Dad must have reasoned that Alice, at age eight, and I, at five, did not need to know. Only by a change in the air, by the emotional straightjacket that seemed to have been put on our family, did my sister and I divine that "something bad had happened." Today, from the perspective of an adult, one who has spent her life working with families that have been blindsided by the intrusion of a fatal hereditary disease, I can only imagine the minefield my parents were navigating at the time.

In school, we had drills to practice protecting ourselves should the atomic bomb be dropped on us. But there were no rehearsals for the bomb dropped on our family by the revelation of Huntington's truths. And the fallout was compounded by the web of mystery surrounding it. My sister and I were aware only by proxy that we had been hit.

One thing is certain. Whatever my father may have believed, and when, he tried his best for a long time to put Huntington's chorea out of his mind, focusing instead on the financial challenge of trying to help Mom's brothers. He believed he would be responsible for supporting them when they were no longer able to work. That conviction, he admitted, helped turn his attention toward matters within his purview, just when his world seemed to be spiraling beyond his control.

♦ ♦ ♦

In the wake of the news about Mom's brothers, Dad, with the deftness of a magician, produced a distraction: a family trip to Norway on an ocean liner, the *Stavangerfjord.* A Norwegian psychoanalyst he met at Menninger's had invited him to Oslo to treat a hospitalized patient with schizophrenia and Dad decided we should all go with him. He would spend the summer working with his patient. Meanwhile Mom, Alice, and I would be delightfully distracted. And we would all be together. After Oslo, he told us, we would move to California, his birthplace that he had dreamed of returning to one day.

And so it was that on my sixth birthday, July 19, 1951, aboard the *Stavengerfjord,* the Captain crowned me Princess of the Ship. My "crown" came in the form of a big red satin bow for my wrist that I became quite attached to. I might be wearing it to this day if my sister, jealous of the attention-hogging Princess Nancy, had not torn it to pieces.

*Princess of the Ship, 1951.*

Six years earlier, a blonde, blue-eyed crowd pleaser had been dropped into Alice's life, and she had not taken her displacement from only-child status easily. After a failed attempt at poisoning by feeding me grass through the bars of my playpen, she accepted that I was there to stay and recalibrated our relationship to her advantage. I was a captive audience, and Alice was running the show.

Eventually, though, we became a great team. We shared very active imaginations. Alice would regularly invent a new universe for us to inhabit, and without hesitation, I would enter it with complete conviction. Perhaps her greatest creation was "Princess Sunset." She had a special voice for the princess that cowed her subject into instant submission. "Princess Sunset orders you to do the dishes!" Her wish was my command. One day, I emerged from the spell she'd cast: "Wait a minute ... you're Princess Sunset! All this time you've been her!" Her transformation into the princess was so complete that it took me weeks to make the connection. I would not be so gullible after that.

In the aftermath of the snow globe toppling, the imaginary worlds Alice created provided a bubble of protection against the disarray going on outside. In her worlds, we always had each other while Mom and Dad were otherwise engaged.

♦ ♦ ♦

Dad had good reasons to surmise that financial responsibility for his wife's family would fall on him. Mom's eldest brother, Jesse, made a modest living as a salesman; Paul and Seymour lived the financially precarious lives of professional musicians. None of the brothers had children, possibly because of warnings against passing on the disease. And now, their days as entertainers were numbered. Although Dad hated leaving Menninger's, salaries there were low and he knew he could earn more income in private practice in Los Angeles. In retrospect, I think this was not an entirely altruistic act as leaving Kansas opened a path to the larger life he envisioned as well.

For Mom, too, a wife and mother with a successful, charismatic husband and two high-achieving children, moving to California was in some ways part of the American Dream. And yet she did not have the professional connections and projects that would enrich Dad's life. After adjusting to Topeka and making a new community there, she was moving once more, this time even farther away from her family and old friends in New York.

Distraction was once again her friend as she watched Dad's career in Los Angeles take off like a rocket while she cheered him from the launch site. But as she settled into a new life in LA, the reality of her brothers' situations must have begun to weigh heavily, her sorrow perhaps exacerbated by distance and her inability to visit them often on the East Coast.

And so, as we adjusted to our new California lives, Mom began the withdrawal into depression that would make it increasingly difficult for her to show up for the life she loved. She didn't go down without a fight: Her depression didn't stop her from becoming the leader of my Girl Scout troop. It didn't stop her from becoming a watercolor painter, or from protesting anti-Semitism at the Riviera Country Club down the street from our house. It didn't stop her from earning a junior college teaching credential in biology in 1967, although she'd been out of the classroom for thirty years, or from inspiring her daughters to become independent and free-thinking women. But it did make her increasingly less dependable. We never knew whom we would wake up to that day. Would it be our mother or someone resembling her but not really her, as though she had sent out her avatar?

Dad was preoccupied with his burgeoning career. He'd chosen to come to Los Angeles with the encouragement of an influential colleague at Menninger's, David Rapaport, a man he revered, who had provided introductions to analysts on the West Coast. Once we arrived, Dad began making friends, and not only with members of the Los Angeles Psychoanalytic Society and Institute, or LAPSI, who welcomed him immediately. His social life soon extended beyond the circle of analysts, to actors, artists, architects, musicians, and writers. He blossomed among them. He had found his tribe. Mom made friends too, with the wives of the analysts and parents of Alice's and my friends. But like many housewives of her generation, especially those with young children, she spent long hours isolated at home with no one to talk to, especially about Huntington's, and about her beloved brothers, so far away on the opposite side of the country.

If that wasn't enough, Dad's professional relationship with a female colleague had evolved into a personal one. Maryline Barnard, a native Southern Californian, had also been a psychologist on the staff of Menninger's, specializing in diagnostic testing. She had begun working with Dad in Topeka on his schizophrenia research case. She also became a family friend, though it wasn't until we all moved to Los Angeles in the fall of 1951 that we began seeing much of her, especially after she moved into adjoining offices with Dad and Romi Greenson. No doubt she moved west partly to be closer to her own family in Oceanside, two hours' drive south from LA, but the move also enabled her to continue her professional and personal relationship with my father. The transition was seamless. Even welcomed. Maryline adored my mother and respected her advanced education and accomplishments. To Mom, Maryline was a welcome companion, a familiar face from Topeka days. At the same time, Maryline adored Alice and me, quietly filling our need for a surrogate mother without ever displacing Mom. I can't say what toll sharing my father with another woman took on my mother. I only know Maryline made my father happier, and the fallout from that benefited us all. And to this day, to anyone who questions the arrangement, my defense is that Maryline never deserted my mother. As Mom's life unraveled and everyone dropped away until even her closest friends had abandoned her, Maryline was always there. For a very long time, it was down to Dad, Alice, me, and this complicated "other woman." She stuck by my mother to the very end.

*At Lake Tahoe, ca. 1958. (*Left to right*) Mom, Hildi Greenson, Romi Greenson, Maryline, me, and Alice.*

So, by the time I was the water-skiing teenager on Lake Tahoe in the late 1950s, the dynamic I've described had settled into a way of life. Celebration had a tinge of sadness. And the patina of privilege was tarnished by curse. When I was fifteen, I went to the movies with Marilyn Monroe, a memorable once-in-a-lifetime event. As we left the theater and walked down the street she demonstrated how walking one way—as Marilyn the movie star—made heads turn. When she walked another way, trying to go unnoticed, she could fade the technicolor version of herself and appear as a regular, albeit extraordinarily attractive, woman. I saw how she created a bubble of privacy around us, a lesson I tried to replicate years later when I became conscious of people watching me. Strolling along, we were just Marilyn and Nancy. She was natural, unassuming, down to earth, and also funny. She made me feel like a friend rather than a fan. Two years later, she would die of an accidental overdose at the age of thirty-six.

In the fall of 1959, Alice left for Palo Alto, starting an independent life as an undergraduate at Stanford. Dad, too, was often away from home at conferences and seminars or giving lectures or attending evening meetings. Or he was holed up in his study, writing. And I was alone with Mom.

♦ ♦ ♦

"I love you, but I no longer love your mother."

The shock I felt that spring of 1962 when my father broached the idea of separating from Mom wasn't as much from what he'd said but the fact he'd said it. Out loud. Now there was no un-saying it. I remember pleading with him, can't you stay with her? We need you. We love you. I love you. His argument was simultaneously considerate and self-serving. He reasoned that she would be happier with someone else and that his staying would deny her the opportunity to meet that person and move on with the better life she deserved. And the sooner he moved out, the sooner she could begin.

As a child I knew everything, and I knew nothing. I experienced the essence of things before I had thoughts or words or facts to attach them to. I knew my mother held something inside her before I knew to call it a "secret." I felt the tension between my parents before I understood what "tension" was. I simply felt what they were feeling. And when they felt bad, I felt bad. So, in a very self-serving way, I made it my mission to make them feel good, to try to bridge the gulf between them. They were, after all, my whole universe at the time. Dad wanted the woman he had married, and she was increasingly unavailable. I didn't understand what was making my mother's light dim. I only knew it was up to me to make it brighten.

This therapeutic attitude of mine began early in my life, perhaps on the playground at UES—University Elementary School as it was called then—the progressive teacher-training "lab" school on the UCLA campus in Westwood that Alice and I attended in the early 1950s. The school mainstreamed kids with disabilities but acceptance into the school did not guarantee acceptance by their peers. And so several of my disabled classmates became my recess therapy group. Getting to know them and seeing how others treated them was my introduction to an attitude that I would fight against for the rest of my life: "Disabled is less than." I find among even some of the most enlightened and progressive people an appalling prejudice against anyone who displays the slightest hiccup in so-called normality. My therapy amounted to simply treating them as the equals that they were.

On a different note, when my uncle Seymour visited soon after we arrived in California, I delighted at his coin tricks and his clarinet and saxophone playing. If he seemed a little jiggly, I barely noticed. But on later visits and when we saw him in New York, he would drop the coins and his music strayed off-key. Mom would freeze and pretend nothing was wrong, while I would leap up to hug him and wrap him in praise and approval. I didn't know why he seemed so clumsy. No one said anything about it, although Mom's reaction must have raised some kind of concern. And it was up to me to make everyone feel okay. I was fixing the world any chance I got. Because if I fixed the world, I could fix my mother. Fix Mom and Dad.

Empathy is not for the weak. I believe I inherited empathy and strength from both my parents. You could say I was, in the language of genetics, a double-dose individual. But empathy needs boundaries, and they are learned. Dad didn't need me to be who he was. Mom didn't need me either. Until she did. During my last two years in high school, when Alice was away in Palo Alto and Dad still lived with Mom and me at the Napoli Drive house but wished he didn't, her need grew exponentially. The comfort of a hug when her sadness made it too hard to speak evolved into my helping her get dressed, helping her make dinner. It wasn't just helping her decide what clothes to wear but helping her put them on. Not just helping her by making the salad while she stirred the pot but helping her put the tasks involved in the right order and then helping her perform them, one by one.

Years later, Dad said he thought at the time that Mom was immersed in grief over her brothers' long illnesses and deaths—Jesse died in 1960 and Paul in 1962; Seymour would pass away three years later. All three brothers gone within five years. Each time she started to regain her bearings after one loss, the next one died. She traveled east for each of their funerals and returned home sadder and more debilitated each time. Surely such a steady onslaught of deaths could have felled someone with many more resources than Mom. Dad tried to help, but he was busy. I doubt he was paying much attention. His thoughts were elsewhere.

In retrospect, I can see that Mom wasn't simply mourning her brothers or anxious about herself. Something else was wrong. At the age of fifty-four, she needed more and more help. Not only a housekeeper to clean once a week, which she had; not only a therapist to talk to once a week, which she had; but medical help as well as assistance with everyday tasks at home. But to bring in that person would have been tantamount to an admission neither of my parents were prepared to make. As long as I was there, it was just Nancy helping Mom.

Life at home and life outside grew increasingly separate. I began hanging out with my best friend, Annie Widmark, daughter of actor Richard Widmark. Their house became a respite from mine, a home away from home, with its carefree atmosphere, gatherings of friends, and first-run movie showings on the backyard patio. And I did well in school. In middle school, in the winter of 1960, I won an "Americanism" essay contest sponsored by the local Pacific Palisades Women's Club, writing about "democracy." In it I noted earnestly—and presciently—that "I cherish the opportunity to take a hand in the government I am living under."[2]

The following year, as a high school freshman, I won an American Friends Service Committee scholarship, enabling me to spend a summer planting trees on farms in France and in Finland, my first journey abroad since our family sojourn in Norway. Two exchange students had come to live with us for a year, I told the local newspaper, inspiring in me a desire "to learn of other peoples ... and to show them America through myself."[3] A desire that would later be fulfilled in ways I could hardly have imagined at the time.

I was not aware of trying to keep my mother a secret. A secret is something that has to be guarded. My home life was simply separate, not unlike the way people separate their professional and personal lives. I knew my teachers as they existed in school. I really didn't think of them outside that role. It seemed perfectly natural.

During all this time, I was in thrall to an obsession, always a state of grace for me. (I think we owe the progress of humanity to obsession.) My obsession was a high school science experiment I had devised, with the help and influence of my father, to examine the potential of affection in ameliorating trauma. I had two groups of mice and a large gong.

Friday, Jan. 29, 1960-Page 7 THE PALISADIAN

# Nancy Wexler Author Of Winning Essay For Americanism Contest

*We are happy to reprint this week the essay by Nancy Wexler, daughter of Dr. and Mrs. Milton Wexler, 1124 Napoli Dr., that was selected by the judges of Americanism Essay Contest sponsored by the Pacific Palisades Women's Club. Nancy received a $10 award and her essay is now entered in the Marina District competition. Nancy is a B9 student at Paul Revere Junior High School – Editor.*

## THE EVOLUTION OF DEMOCRACY

What is democracy? Most people will call the United States a democracy; most people will consider Russia undemocratic. But is this an answer? I know that I should be thinking noble thoughts of democracy every time I stand in honor of our flag. But I confess I can't because I do not really know what a democracy is. I do not know what the idea of democracy is. I think it must be something very intangible and elusive, yet something of great force which has served to strengthen our American freedom. In this essay I should like to answer my own question on what a democracy really is, both as an idea and as a material fact.

I feel that in order for me to understand this thing called democracy I must know from whence it sprang, its birth and evolution. Ever since history was created man has lived together in clans or tribes which gradually expanded into larger and larger organization, villages, cities, states, then vast countries under a single government's control. With the growth of man's civiliation, man's mind too, has expanded until it could grasp the concept of a democracy, a rule "of the people, by the people and for the people."

The first seeds of democracy appeared 450 years before the existance of Christ, when Pericles, an early Grecian ruler, was in power. This man made it possible that all Athenian males who were free citizens could participate in the transactions of the government. Compared to the kings, rich men, and tyrant rulers of before, Grecian democracy was a radical step in the right direction.

Democracy had only a beginning however. For many centuries, nearly 2,000 years in fact, it lay dormant. In 1066 William of Normandy started his Norman Great Council, which served as a check on the selfish ambitions of future kings. In this early organization the germ of democracy can be detected. One of the first kings to feel the council's force was King John of England. When he signed the Magna Carta in 1215 Democracy had established a foothold. In this document King John conceded the freedom of the Church, certain rights of noble men, and abolished certain taxes. The Petition of Rights and the Bill of Rights were also important measures toward a completely democratic government. Democracy was firmly established, if not as a function, then as an idea.

During those troubled times of England, the Declaration of Independence had severed United States from our mother country, England, and we were left to our own devices to establish a democratic government. The Articles of Confederation were a weak attempt at establishing a government of this sort, but the people clung too closely to their individual freedoms They were afraid to give central government the power it needed to efficiently control, and the articles failed in their endeavor. With the coming of the Constitition democracy was truly established in the United States. An intricate system of checks and balances of the judicial, legislative, and executive branches of the government was set up. However there was something still missing. still a gap to be filled. The Bill of Rights had to be written to guarantee Americans the freedom to think, to act, to believe, to speak, write, hear, worship, assemble, and possess; these were the basic principles and supreme functions of democracy.

I am still unsatisfied. A background knowledge does not answer for me adequately the question of what is democracy. To different people a democracy represents different things. To Socrates, that renowned Greek philosopher, a democracy was a place where political abuse inevitably prepare the way for despotism, or rule by a dictator. But is this what we have today? Was Socrates right? I am inclined to say no. My logic tells me this is not so. I deeply value the principal of government by the people themselves. I cherish the opportunity to take a hand in the government I am living under. The idea the belief in, the practice of social equality, and the disregard of social barriers based on class, birth, race, etc., a definition of democracy on a more philosophical level, is one that greatly appeals to me. One might even say that democracy is a way of life which brings increasing opportunities of development to all people in every walk of life.

Just a glance at the United States' history is a convincing argument against Socrates' prediction. Look at all the heroes who have died for the sake of democracy and

PRESENTING THE AWARD for the winning essay in the Americanism Contest on Democracy sponsored by the Pacific Palisades Women's Club, Mrs. Norman Loretz, Americansim chairman of the club, gives a $10 check to Nancy Wexler (fourth from left). Others in group are: (l-r) Dr. and Mrs. Milton Wexler, Mrs. Sylvin Bilsky, Deborah Bilsky, Mr. Bilsky and Mrs. H.W. Linton. Deborah was author of the second place essay.

I think it is only fitting that I, in closing, quote one of the leading men of all times, whose simple definition epitomizes the very ideal of democracy and makes all other explanations unnecessary. "We hold these truths to be self evident, that all men are created equal, that they are endowed by their creator with certain inalienable rights, and that among these are Life, Liberty, and the Pursuit of Happiness." Thomas Jefferson.

*"Defining Democracy." (Reprinted, with permission,* The Palisadian Post, *January 29, 1960, p. 7.)*

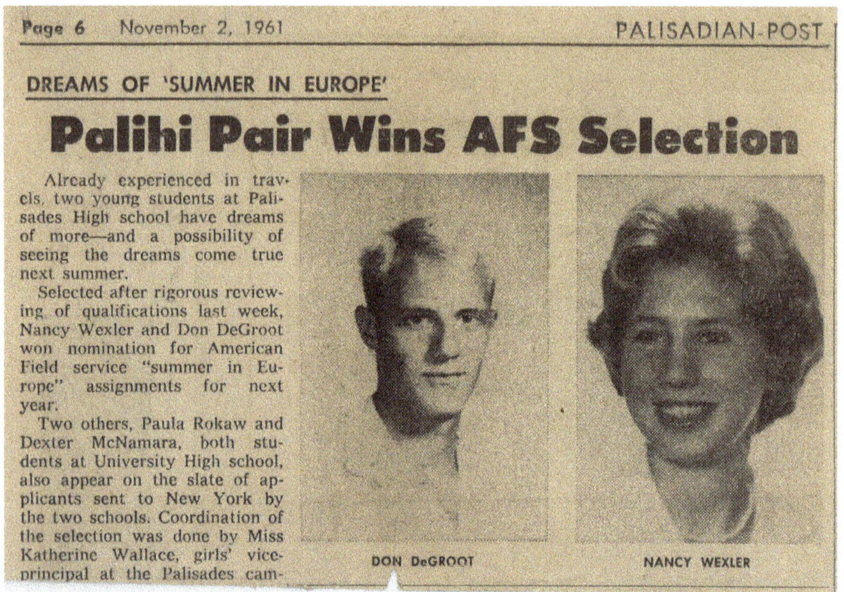

Page 6 November 2, 1961 PALISADIAN-POST

DREAMS OF 'SUMMER IN EUROPE'

**Palihi Pair Wins AFS Selection**

Already experienced in travels, two young students at Palisades High school have dreams of more—and a possibility of seeing the dreams come true next summer.

Selected after rigorous reviewing of qualifications last week, Nancy Wexler and Don DeGroot won nomination for American Field service "summer in Europe" assignments for next year.

Two others, Paula Rokaw and Dexter McNamara, both students at University High school, also appear on the slate of applicants sent to New York by the two schools. Coordination of the selection was done by Miss Katherine Wallace, girls' vice-principal at the Palisades cam-

DON DeGROOT

NANCY WEXLER

*Summer in Europe. (Reprinted, with permission,* The Palisadian Post, *November 2, 1961, p. 6.)*

Regularly, I would ring the gong close to the mice and traumatize them with the sound. Then, with the first group, I would take a mouse in my hand, stroke it, and speak softly to it, bathing it in affection while carefully charting its physical reactions, such as how quickly its heartbeat returned to normal. The second group, my control group, was subjected to the same trauma but given no follow-up affection. I was only a few days into my experiment when one of the traumatized mice had a heart attack and died in my hand. My father came home to find me face down on the shag rug sobbing and wailing, "I'm a mouse murderer!" To which, smiling, he shook his head and said, "Nancy, you'll never be a scientist." As the memories of the two incidents juxtapose—my father's pronouncement that he no longer loved my mother and the death of the mouse—I think now that I was sobbing for the failure of a grander experiment.

♦ ♦ ♦

The summer of my high school graduation in 1963, Dad moved into his own apartment, and I prepared to leave for Radcliffe, the sister

*With a teacher and classmates on graduation day at Palisades High School, June 1963.*

college of Harvard, three thousand miles away from home. While this turning point in most young lives represented the child separating from the parent, I was in the unique position of having become the parent to my own parent. I worried about how Mom would fare without me living in the house with her. And yet, even though I felt ambivalent about leaving her on her own, I knew I had to leave home to save myself. I resolved to write to her and call as often as I could, to take care of her long distance.

For all her neediness, Mom always championed me flying far and high. She was thrilled that I had been accepted at Radcliffe. For her, the East Coast was the bastion of higher learning. She had done exceptionally well at Hunter and Columbia, and she wanted me to do the same. In her send-off I felt the unspoken charge to go out in the world and do all the things she hadn't done, supported by her love and confidence in me that never wavered, no matter her neediness or absence at times.

I had come to accept that life isn't fair. But I never thought it could be as mocking and sadistic as it seemed that summer of 1963, before I left for Radcliffe, when I joined Alice in Guadalajara, Mexico, to take a six-week Spanish class. I was pleased when Mom decided to visit us, her first journey on her own since she and Dad had separated. I was looking forward to spending time with her and showing off my Spanish. On her way, she stopped over in Mexico City to do some sightseeing. One night, as she returned to her hotel after dinner, four young men jumped her taxi, seizing her wallet, earrings, and wedding ring. They threw the driver out, commandeered the cab, and drove into a remote neighborhood where they dragged her out of the car to take turns raping her. Then they left her there alone, beaten and bleeding. She stumbled from house to house until someone finally helped her get to a hospital, where she got treatment for her injuries and called my father. He immediately flew down to bring her home.

It seemed as if the universe were sending my mother a cosmic message: "You don't deserve a life." She took a leave of absence from the classes she had begun at UCLA and sank further into depression, now exacerbated by anxiety attacks and an obsession with locking all doors. For quite some time, I'd had the growing sense that Mom was up against insurmountable odds as I watched her emotional equilibrium repeatedly undermined. Every time she gained some stability, life stuck its foot out to trip her. And each time she stumbled, it was harder for her to regain her footing. I began to doubt she ever would.

It was following Mom's third trip east, in late March 1965, to bury Seymour, the last of her brothers to die, that Dad took me out for dinner at The Luau, a Polynesian extravaganza on Rodeo Drive in Beverly Hills. I had come home from Radcliffe for spring break, knowing that my mother was now living alone in her own apartment. (Alice was in Venezuela on a Fulbright fellowship for the year, knowing nothing about Huntington's.) By this time, Mom and Dad were divorced and the family house had been sold. I remember that I wore a new black-and-white tiger print silk shirt I'd chosen just for the occasion. As Dad and I entered the restaurant, crossing a drawbridge over a moat, passing statues of Tiki gods, I thought maybe this festive night out was his way of thanking me for giving him his freedom by taking so much responsibility for Mom.

"Nancy, I need to tell you something." His tone did not match the décor. "Your Uncle Seymour, your Uncle Paul, and your Uncle Jesse all had the same thing. It's called Huntington's chorea."

I wasn't ready for such a serious discussion. A waiter passed by, delivering a flaming drink.

"Huntington's? They're Jewish. How could they get something with such a *goyishe* name?!"

He was not amused. "It's an inherited disease. Your grandfather had it as well."

Inherited. A sobering word.

"Can Mom get it?" I asked.

"No. Because women can't get it. So, you're not going to get it and Alice is not going to get it."

He knew at the time that this was not true. He simply lied. "To protect you," he later said. And I thanked him for it. I wouldn't have been able to handle the truth at that time, at the age of nineteen. It's hard enough to handle it now.

But as I think back to that memorable night more than half a century ago, I ask myself whether my father may have communicated more to me than he had intended. Could he have conveyed his fear, his knowledge—perhaps by the intonation of his voice, the cadence of his speech—that Mom was vulnerable too? And that Alice and I were at risk as well? Did he insist too much that "women don't get it"? For as I left The Luau that night, stepping out of the Polynesian fantasy, I felt myself crossing over the drawbridge into a new reality. An unfamiliar word, "Huntington's," had entered my consciousness though, for the time being, I remained protected from its full significance. Any subliminal messages I was receiving from my father remained submerged.

♦ ♦ ♦

Whatever guilt I felt about being at Radcliffe and leaving Mom alone was allayed as she slowly took up painting again, invited a few friends for dinner, and even returned to UCLA to complete the requirements for her teaching credential. However, any vulnerability is exacerbated by trauma, and my mother had little margin for error. The ensuing several years marked a decline from which she would not rebound. While her hold on life grew more tenuous, mine was expanding. At Radcliffe, I made new friends and immersed myself in classes with a host of eminent and engaging professors, including the psychologist Erik Erikson, who became my advisor and mentor. I double-majored in social relations (known as SocRel) and English, writing my honors thesis on Mary

Ann Evans, aka the novelist George Eliot, a project I loved, except for the deadlines which I always managed to make but just barely, sometimes thanks to the help of my visiting father, who volunteered to type my pages at the last minute.

I was thrilled when I received a Fulbright fellowship for the year following my graduation, just as Alice had several years earlier; the Fulbright Program sent me to Jamaica, where I enrolled as a graduate student at the University of the West Indies in Kingston. Like Alice in Venezuela, though, I ended up spending much of my time outside of my classes, with a welcoming Jamaican family and with my new boyfriend, Michael Collins, an adventurous redheaded Peace Corps volunteer and future filmmaker who remains a lifelong friend.

Later on, to accommodate visitors, including both Mom and Dad, I rented a small house with a circular driveway and high windows open to the soft Caribbean breezes. Visiting me in January 1968, Mom shared my bed. Falling asleep was difficult for me because Mom was so restless, seemingly unable to get comfortable until finally she dropped off to sleep herself. I attributed her twitchiness to a reverberation of her experience in Mexico. She was nervous in a strange place, uncomfortable being in an unfamiliar country.

*With Michael Collins in Jamaica. (Photo by Michael Collins.)*

Reports came back from Alice when she later visited Mom back home: Mom reciting a litany of regrets for the times she had been unavailable to us; how her depression and fear had made her unable to hear us, and that now that she felt able to, we were no longer there. The plans for returning to teaching, along with her other projects, seemed to have evaporated. Nervous, anxious, and argumentative, Mom's energy had become focused on shopping for clothes.

♦ ♦ ♦

That spring of 1968 I got a phone call from Dad asking me to return to Los Angeles for our annual summer visit in time for his birthday in August, somewhat earlier than we usually visited. I should have known something was up. He wasn't sentimental about his birthday, and he'd never celebrated them with any fanfare. But this was his sixtieth, not just any birthday. He'd already called Alice, he said, and she 'd agreed to come. At the time, I was in Europe, having left Jamaica to spend several months observing at the psychoanalytically oriented Hampstead Child Therapy Clinic in London, run by Freud's famous daughter Anna and her partner, Dorothy Burlingham. But I was ready to come home.

Amazingly, Alice and I arrived at the LA airport within twenty minutes of each other. Dad picked us up—unusual, since typically he and Mom came together, even after their divorce. "Where's Mom?" I asked. He told me Mom was sick. "Yes, I know," I said, surprising myself. I wasn't thinking of Huntington's; I thought she was psychologically sick. We drove to Dad's apartment, where, for some reason, we went into his bedroom. Alice and I sat down side by side on the edge of the bed, the way we used to do as kids. We sat facing a huge painting of Humpty Dumpty falling off the wall but grinning desperately while our father told us the real reason he had called us home: "Your mother has been diagnosed with Huntington's chorea."

The chronology of the conversation is a blur to me now. I no longer recall whether he explained why he had withheld the truth about the inheritance of Huntington's, or whether he mentioned that it was always fatal, before or after he told us about Mom's diagnosis. I do know what he said last: "Each of you has a 50–50 chance of inheriting this yourself." Alice responded without missing a beat: "50–50? That's not so bad." Somehow, she had thrown us all a lifeline. It gave comfort to Dad and gave me time to think. My first thoughts were that my mother was going to die and that we should not have children. Third in line was that we could get this too. I remember that being the order of my concerns.

Dad told us there was no cure for Huntington's. But at the same time as he was telling us Mom's diagnosis on that sultry August afternoon in his bedroom, he told us about his meetings with local neurologists and geneticists and about the New York–based Committee to Combat Huntington's Disease, or CCHD as everyone called it. CCHD was the first ever Huntington's advocacy group, organized by Marjorie Guthrie, widow of folk icon Woody Guthrie, who had Huntington's and had died the year before. Dad had already begun organizing a California CCHD chapter and was making plans to raise money for research.

I needed time to absorb all this information. I needed to reconfigure my understanding of my mother's life, now that the pieces were starting to fit together. I needed time to mourn and to reconsider what our future lives might be, Alice's and mine. But of one thing I felt certain: If anyone thought that Mom's life was a prophesy of what mine would become, I was going to prove them wrong.

CHAPTER

2

# IMPLOSION THERAPY

The day after the revelation in Dad's bedroom, I went to see Mom in her new apartment, the first time I'd seen her since she had visited me in Jamaica six months earlier. I was relieved to find her essentially unchanged. The tiny twitches of her fingers and toes and her slightly unsteady walk were no more pronounced than they had been before. But now a cause had been identified and named, and I was seeing her through that lens, the Huntington's lens.

"Your father told you about my illness?"

Right to it. "Yes. Huntington's disease."

"Oh no! Please don't make it any worse than it is. He said the diagnosis is 'demyelination of the nerves.' It's like multiple sclerosis. Not a good prognosis for me," she continued, "but for you and Alice, it's a godsend. It's not my brothers' illness. Nor my father's. I could not have passed it on to you. I can live with this; I could not live with that."

I couldn't fault Dad for orchestrating the lie. Hadn't she just told me the truth would destroy her? Mom told me how she had been on her way to jury duty. She'd driven downtown, had parked her car, and was walking toward the courthouse when her weaving walk caught a policeman's eye. He stopped her and berated her for being drunk, and so early in the day. She called my father in a panic. He made a doctor's appointment for that afternoon, and then she continued to the courthouse.

The emotion I felt as I listened wasn't only pain from hearing the story of the day's events, it was also admiration, even awe. My mother, who had endured a tsunami of family deaths, a gang rape, her husband leaving her, and her children leaving home, got up that morning and through the fog of her depression and grief drove herself downtown to do her civic duty. To be in the world. And now she was sitting in front of me dying. Or so I thought at the time. I was in free fall and the ground was coming up fast.

My first impulse was to say the kinds of things I would hate hearing years later: "Look how far you've come in life without it affecting you! Why should it stop you now? Look at everything you're accomplishing!" Rushing in to comfort. It's human nature. But the urge to comfort is often borne out of selfishness. We are all trying to help—ourselves. My best thinking told me to say nothing. I took her in my arms and hugged her.

I am a big believer in the healing power of the hug. "Nancy is a hugger" often precedes me, almost as a warning. In all those moments when it's best to say nothing, a hug is a way to say everything. In this case, it was also a way to avoid furthering the fabrication. Maybe my mother's emotional state *was* like a house of cards that would topple under the weight of the truth. Or maybe she knew the truth and found it too painful to accept. My parents shared a prodigious capacity for denial. Together, they had an almost savant ability to hold the truth about bad things at bay—from their children, from each other, from themselves. Or perhaps they had tacitly agreed to a diagnosis that would make life more livable for everyone concerned. I could play that game too. I had studied at the feet of masters. At the same time, I had learned to channel denial to higher purpose. Denying impossibility. Denying hopelessness. Denying the odds. Over time, sustaining the myth that Mom did not have Huntington's began to feel less like denial in the service of kindness and more like lying to my mother.

♦ ♦ ♦

I was about to start a PhD program in psychology at the University of Michigan, in one of the few psychology departments in the country with a clinical and specifically psychoanalytic orientation. One of Dad's trainees from Menninger's, Martin Mayman, was on the faculty, and I looked forward to working with him. Even before Mom's diagnosis, Dad had thought up the idea of a road trip that summer: We would celebrate his sixtieth birthday in LA, after which he would drive Alice and me east across the country. We would drop Alice off in Bloomington at Indiana University where she was working on her PhD in history and continue to my destination in Ann Arbor. Dad thought it best to keep to this plan. All three of us hovering over Mom could have raised suspicions in her mind. Autonomy and independence were the best medicine. And she already had full-time live-in help; in truth, a caregiver, but someone carefully cast by Dad to play the role of "housekeeper."

We said our goodbyes and headed out of LA on the old Highway 66, Dad at the wheel declaring, "We're gonna lick this thing," as he tried to reassure both himself and me and Alice. Once we hit the road, I could feel his bravado infecting me. "We're going to find the cure!" I announced, riding shotgun. Alice sat in the back seat, resigned to our giddy optimism, the lone member of the trio brave enough, or skeptical enough, to sit quietly with the feelings we were shouting to drown out.

Driving north through Las Vegas and on to Salt Lake, hypnotized by the open road, I felt a swelling confidence. As we followed the Snake River into the expanse of the Grand Tetons, I had the revelation that it was easier to be optimistic about Mom's prognosis when she wasn't sitting in front of me. I understood why a surgeon never operates on a member of her family.

My mind was opening along with Dad's. We were in Yellowstone National Park standing in the spray of Old Faithful when he proclaimed, "The genetic revolution has begun, and anything is possible! All we have to do is believe!" I took Old Faithful's shooting 100 feet into the air as nature giving him a thumbs-up. "I believe!" I shouted, laughing.

The physical distance between me and my mother had momentarily freed me to start absorbing what had happened, though I still had only a vague idea of what Huntington's actually meant. We delivered Alice to Bloomington, leaving Dad and me to conspire together as we drove north toward Michigan. I was feeling almost jubilant as Dad and I discussed the future, imagining breakthroughs in genetics that would cure not only Huntington's but many other ills. But the moment he left Ann Arbor, I felt myself in an alien landscape where I was entirely alone. As classes began, unwilling or unable to process the issue of Huntington's, I assumed the identity of a person free of problems. I entered the emotional equivalent of the Witness Protection Program.

I kept in touch with Mom through letters and phone calls—all about me, exactly what she wanted: my starting out as a psych grad student; rushing between classes, case presentations, and advisor meetings; making new friends. The whirl of a life opening up provided the perfect antidote to a life closing down. Despite my discomfort with our masquerade, we kept up the fiction about multiple sclerosis, with occasional attempts by me to broach the truth.

The following summer, I saw an opening. I decided to test the waters. The feature film *Alice's Restaurant,* starring singer-songwriter Arlo Guthrie, son of Woody Guthrie, was released in theaters, and reviews indicated that Woody was a character in the film. I would take Mom to see it as a way of saying, "Look, we can talk about this." I wanted to tell her about CCHD and Dad organizing the California chapter and already raising money for research. By this time, one year after Mom's encounter with the police officer, I too had met Marjorie Guthrie as she traveled around the country encouraging HD family members, including me, to come out of hiding and organize chapters in our regions. I had begun meeting with some of the families with Huntington's in Detroit who were organizing a Michigan chapter. I imagined telling Mom about all this after we saw the film and giving her the opportunity—completely uncoached—to broach the subject of Huntington's. In any case, I would follow her lead and see where it took us.

On the screen, Arlo, playing himself, is visiting his father, now hospitalized in a late stage of Huntington's and barely able to move on his own. I had never seen a person in this state, so it took quite an effort on my part to stare straight ahead, artfully dispassionate, as if this were just another scene in the movie. In truth, I found the scene both scary and oddly reassuring. The Woody figure on the screen (an actor, of course) looked quite healthy—no wasting, no pallor of the terminally ill. He was, however, flat on his back, unable to speak, unable to move. One hand lay lifelessly across his chest, as if someone else had made the decision to put it there. In the scene, Arlo watches as Marjorie carefully places a small piece of cookie in her husband's mouth. Chewing and swallowing seem to require his full concentration. She then lights a cigarette. We discover that it's for Woody, as she puts it between his lips, leaving it to hang out of the corner of his mouth as he puffs. When Arlo is ready to leave, he bends over his father and kisses him on the forehead. Woody acknowledges the kiss by blinking his eyes.

Sitting next to Mom in the theater, I felt the movement of her agitation but ignored it to allow her privacy. And then her voice in my ear insisted that we leave the theater. Not waiting for me, she rose from her seat and started up the aisle, as though someone had yelled "Fire!" Outside, she refused to talk about what had upset her. With the hubris of youth, I overlooked the fact that, for my mother, seeing anyone with Huntington's would likely have brought up painful memories. Maybe she looked at Woody onscreen and revisited her father Abraham in the back ward of the state hospital. Or replayed scenes from the last years of her brothers' lives. Perhaps she saw herself lying in that bed, the captive audience of a sadistic fortune teller, with someone stuffing a cigarette into the corner of her mouth.

When Mom made it clear she didn't want to talk about the movie, I understood that I'd overstepped unspoken boundaries. Following her lead, we talked about me and my psych classes at the university. Doing so provided a safe area of discussion and, by implication, eliminated any suggestion that Alice or I might alter our trajectories in light of recent developments. Mom and I did not discuss Huntington's for a long time to come.

♦ ♦ ♦

For a while in Ann Arbor, I continued as if Huntington's had not just upended my life. I told no one about my mother. War was raging in Vietnam in the early 1970s, and our graduate class was small, nine or ten students, mostly women. Practical case work began in our second year, and I was assigned to the psych clinic. Once a week, staff members presented a case. One case especially stands out in my memory.

Doris, a staff social worker, was treating a family with profoundly dysfunctional parents and a seriously disturbed ten-year-old son with no clear diagnosis. We discussed whether the boy might have autism or schizophrenia, a learning disability, or some other condition. The mother repeatedly blamed the father for their son's difficulties. At some point, the maternal grandmother, whom Doris described as an alcoholic, died in a psychiatric hospital of cirrhosis of the liver. Later, it turned out that the grandmother also had Huntington's. The mother had never told a soul. She was terrified that *she* might have the disease and that she might have passed it on to her son. For years, she had been living with this devastating secret. After the secret was revealed and the family was able to talk about the disease, their situation got much better. Just being able to speak about it radically improved this family's life.

Doris's case impressed me greatly, first because it was about Huntington's, and second, because it demonstrated the beneficial impact of having the facts out of the closet. For me, that realization was an epiphany. It completely changed what I thought was possible. There was no medical treatment to stop Huntington's but there *was* psychological treatment; talk therapy could help people deal with it. Help Mom deal with it. Help *me* deal with it. I was finding the fiction that Mom did not have Huntington's and that Alice and I were not at risk difficult to live with. I wanted to talk honestly with her about Huntington's. Talking was therapeutic. Talking could help our family too. Doris's case made me realize the importance of being active. I felt empowered by this story, and it made a huge difference in my life.

♦ ♦ ♦

Meanwhile in Los Angeles, Dad had begun holding meetings with local HD families and reaching out to his medical friends and their colleagues to figure out how to move forward. He was thrilled to find anybody in LA who cared about Huntington's and was willing to come to a meeting or donate money or help raise awareness. From the start, he wanted the California CCHD chapter to emphasize support for research—the legacy from his Menninger days. Although he also soon came to realize the tremendous need for accurate, up-to-date information about the disease, for physicians, teachers, social workers, and police, as well as for families, who were also often misinformed.

One of the chapter's first big fundraising efforts was the Woody Guthrie All-Star Tribute Concert, held at the Hollywood Bowl in September 1970, with such icons of the folk world as Pete Seeger, Joan Baez, Odetta, Arlo Guthrie, Country Joe, and Jack Elliott singing Woody's songs under the stars, while Peter Fonda and Will Geer took turns telling Woody's story in his own words. Alice and I flew home for the event. At the last minute, Mom, who had read about Dad's fundraising in the *LA Times* and insisted that she wanted to come, could not bring herself to attend. I understood. I always felt everything she was feeling. And now I felt her presence as I listened to Woody's words:

> *I hate a song that makes you think that you are not any good. I hate a song that makes you think that you are just born to lose. Bound to lose. No good to nobody. No good for nothing. ... I am out to fight those songs to my very last breath of air and my last drop of blood. I am out to sing songs that will prove to you that this is your world and that if it has hit you pretty hard and knocked you for a dozen loops, no matter how hard it's run you down and rolled over you, no matter what color, what size you are, how you are built, I am out to sing the songs that will make you take pride in yourself.*

The day after the concert, Alice, Mom, and I were lounging around Maryline Barnard's swimming pool where we often gathered. Maryline was the only one of us who lived in a house with a backyard, let alone a pool, and we were always welcome. Face to the sun, stretched out poolside on a chaise lounge, Mom looked so youthful and attractive that the three of us could have been taken for sisters. Her symptoms were almost undetectable. I was basking in the afterglow of the hugely successful concert while telling Mom about the groundswell of interest in the research we were already funding.

She brightened. "Oh, Nancy, that's great." At least she wasn't pushing the subject away. Heartened, I went on to report how much money the concert had raised and told her about the California chapter's new research projects.

"Yes, I read about it in the paper this morning."

So, she *was* taking it in!

"It's all terrific," she said.

Alice and I both breathed a sigh of relief.

"But you know I don't have Huntington's disease."

I had just come from Michigan where I'd listened to Doris's case presentation about the emotional costs of secrecy and denial in a family and how confronting the truth had helped them. I thought it would help my family as well. I also suspected that Mom knew the truth.

"Mom...," I began.

Alice shot me a look that said, "Don't." But that was like asking a high diver in mid-dive to jump back onto the board.

"Mom ... you *do* have Huntington's. At first, they thought it was something else, but now the neurologist has told us the diagnosis is definite."

She started to cry but said nothing.

"That's why we're all so on fire about the research. It's because of you!"

She remained silent. When she'd told me initially that she couldn't live with a diagnosis of Huntington's, I knew that one reason was her concern for us, for Alice and me. It was about what *we* could live with. I wanted to reassure her that I *could* live with this because I was trying to change it.

"Mom, this research is all happening because of you!" I told her. "You were a pioneer in your family, in your own research, now you've got to be a pioneer for us. That little 94-pound body of yours has got the answer. I promise you I'm not going to stop until we find the cure."

Of course, I was trying to convince myself as well as her, and it was working, for me at least. Mom said nothing. I hugged her tight.

♦ ♦ ♦

A few months later, I got the call from my father that Mom had tried to commit suicide. I was in disbelief. Dad relayed the details: the "housekeeper" found her ..., empty bottle of pills ..., rushed to the hospital ..., stomach pumped. It all reverberated in the background as my body reeled, mirroring the feelings that had brought her to the decision to end her life. The fortune teller had gotten to her.

*My mother had tried to commit suicide.*

I understood why. But I didn't care. No mother of mine was going to kill herself. All the talk about patient autonomy and rights and human dignity rang like phrases from a remainders bin of self-help books when they were applied to my life and my mother's. I called her up and bawled her out for doing such a stupid thing. "Do you want to end up worse off than you are? Still alive but brain damaged?" Of course, I told her I loved her, and that Alice and I needed our mother. But I wanted to scare her. I needed to stop her from seeing suicide as an option. She promised that she would never try it again, and she never did.

While the momentum of a new life had kept my full acceptance of Mom's condition at bay up to now, it had also shielded me from the reality of being a candidate for inheriting her illness. In the bloom of youth, running at full throttle, I felt it as a preposterous idea. Maybe I had only imagined that I was at risk for this disease. I hated that phrase "at risk." Doesn't simply getting up every day and going out the door put us all at risk?

All my spinning plates of denial came crashing down in that phone call from my father. My mother had tried to kill herself. Her disease was real. The fortune teller had been right about her. And now, I could imagine that fortune teller coming for *me*.

♦ ♦ ♦

By the spring of 1971, with Maryline's help, Dad had gotten through the bureaucracy of establishing the California CCHD chapter, recruiting a few members, raising funds, and defining priorities. He knew from the start that he wanted to focus on basic research aimed at understanding the disease. Eventually, he got a few projects started, including an HD tissue bank for researchers, located at the Veterans' Administration in LA under the supervision of Wallace Tourtellotte, UCLA professor of neurology who had previously created a multiple sclerosis brain bank at the University of Michigan. Dad also organized a small science workshop in Pacific Palisades with several high-powered researchers including Nobel laureate Julius Axelrod from the NIH, and Lasker Award winner Seymour Benzer, a molecular biologist and geneticist from the California Institute of Technology. William Dreyer, a molecular immunologist and pioneer in biotechnology, also came from Cal Tech. They urged Dad to reach out not only to senior scientists and clinicians knowledgeable about Huntington's (of which there were few) but also to young investigators, postdocs, and even graduate students from different scientific fields and get them together for brainstorming sessions. They didn't have to know anything about Huntington's disease (Benzer and Dreyer themselves knew little at first). What mattered was their ability to think creatively, their willingness to share ideas, and, in some cases, their mastery of new technologies. They also recommended an iconoclastic graduate student at Caltech, Ronald Konopka, who with Benzer had just published a landmark paper showing how genes determine circadian rhythms in flies. Dad immediately hired Ron to seek out talented up-and-coming younger

*Seymour Benzer.*

scientists across the country who might be interested in attending future workshops on HD.

Dad was filled with enthusiasm. He had noticed that psychoanalysts and psychologists at professional meetings often dozed off during slide presentations but talked animatedly during the breaks so he designed workshops to be like the breaks. Thus began the small, freewheeling, interdisciplinary workshops that Dad—and I—came to love so much as incubators for new ideas and settings for building a research community. Dad also drew on his experience doing group therapy with artists in LA. As he put it, why shouldn't science be approached in the same creative way that books get written, movies made, buildings designed, and music composed? Just let yourselves go, he would say at

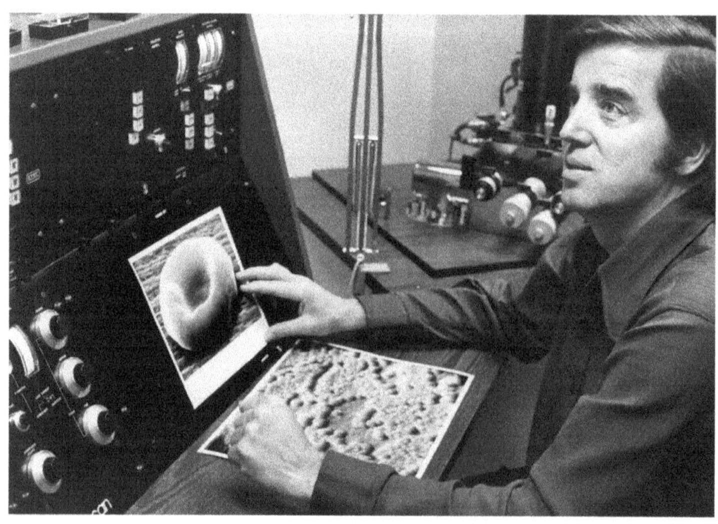

*William Dreyer at Caltech. (Courtesy of Caltech Archives and Special Collections.)*

the opening of a workshop. We want to hear your wildest ideas, so don't be afraid to make a mistake. After all, at this stage, who can tell a mistake from a useful idea? In the beginning, I could feel the young researchers regard this older layman's exhortations with skepticism. But Dad's personal experience with HD as well as his charisma as a psychoanalyst gave him a gravitas and authority that made them willing to listen.

Dad also made the sharing of ideas—meaning unpublished ideas—a requirement of the workshops, another vexing issue among young researchers. At this point, biomedical research was not yet a team sport. A scientist's work is kept alive by grants, and grants are given to people whose discoveries are unique to that person. Think of Thomas Edison showing his plans for the lightbulb to a roomful of other inventors before he secured the patent, and you'll have some idea of what we were asking.

Dad's freewheeling approach to the workshops became his signature contribution. And I take credit for insisting that we needed to give a face to the disease by bringing people living with Huntington's to the meetings. Many of the researchers had never met a person with the disease. They knew them only from blood samples and slides.

"Everyone, meet Anne." Sitting in front of them was a statuesque architecture student and former fashion model in her mid-twenties from Stuttgart, Germany. At six-foot-four, Anne would not fit under anyone's microscope. The room full of young researchers listened like children at story hour as Anne told them about her life, about studying to be an architect like both her parents while supporting herself by modeling, and about playing beach volleyball on the weekends for fun. Despite her symptoms, Huntington's had not yet robbed her of her ability to speak four languages. Granted, she was an exceptionally talented human being, but after meeting Anne, the attendees never again looked at a person with Huntington's in the same way.

♦ ♦ ♦

The workshops soon became the hallmark of the California CCHD chapter and its successor, the Hereditary Disease Foundation and today the Huntington's Disease Foundation. Beyond the workshops, my father also had a great appreciation for one of Hollywood's most enduring maxims: Never underestimate the power of a party—nor of fame. After the first day of that first workshop, the attendees were invited to what became a tradition: the Saturday night party, this time at the art-filled Malibu beach home of actor Jennifer Jones and her new husband, entrepreneur and art collector Norton Simon. Both became longtime supporters of the foundation, with Jennifer bringing glamour—her own and that of her Hollywood friends—to foundation events.

Pathbreaking architect Frank Gehry, who changed the face of twentieth-century architecture, along with his wife Berta, became founding trustees, hosting events and later establishing a coveted foundation prize for innovation in science in memory of Frank's artist daughter Leslie who, tragically, died in 2008 of uterine cancer at the age of fifty-four. Beyond these contributions, Frank's radical approach to art and architecture and the many conversations we had over the years helped shape my father's and my vision, not only for the foundation but for our entire lives.

Dad's painter, sculptor, and ceramicist friends, including Larry Bell, Billy Al Bengston, Tony Berlant, Ron Davis, Ed Moses, and Ken Price, also contributed their artists' cachet to the events they attended, many of them donating prints, sculptures, and drawings to raise funds. And when actor Julie Andrews and her husband, director Blake Edwards, welcomed guests at their Malibu estate, awestruck young scientists could hardly believe they were in the company of *My Fair Lady* and the author of the *Pink Panther*! They *always* recalled these parties and the actors they met, such as Sally Kellerman, Jack Lemmon, and the multitalented Elaine May. Although my memories of those evenings conflate, images float up: future Nobel laureate H. Robert Horvitz describing his research on *C. elegans* (a worm) to Frank Gehry who, sketching as he listened, translated the science into the visual language in which he was fluent; MIT geneticist David Housman explaining the effects of Huntington's on the brain to playwright Lillian

*Architect Frank Gehry and me. (Photo by Michael Collins.)*

*Cleaning lobsters with Alice Pratt for a memorable Saturday night dinner in the home of Ed and Kathryn Kravitz.*

Hellman, her Southern charm and self-assurance fueled by Jack Daniels; future head of the Human Genome Project *and* of the National Institutes of Health Francis Collins trading ideas with performer and writer Carol Burnett, who became a longtime foundation supporter; and Alice Pratt, an early Huntington's advocate whom I would come to know well, cleaning lobsters with me for a dinner at the Cambridge, Massachusetts home of Harvard neuroscientist Edward Kravitz and his wife Kathryn. Ed would become a good friend, playing a vital role, both in the foundation and in teaching the neurobiology of disease.

In tandem with Dad's organizing in Los Angeles and with Marjorie's encouragement from New York, I became vice president of a new Michigan CCHD chapter. I felt like I was living two lives, commuting between my student life in Ann Arbor—as a woman without problems, focused on my psych classes and many new friends who knew little about me—and my CCHD life in Detroit and Los Angeles. The combination was exhausting, and I sometimes nearly fell asleep driving back to Ann Arbor late at night. The truth was, I was depressed and uncertain and a little bit scared. But I was meeting families such as the Wilders (not their real names), a large, loving working-class clan with six children, a mom with undiagnosed Huntington's, and a devoted Dad who nonetheless was overwhelmed with the chaos and at the end of his rope, until the diagnosis identified the problem. I got to know them all and took the kids under my wing; they became my adoptive family in a way, and they helped motivate me and keep me focused. These were folks who valiantly stared down the truth of their lives and carried on. I couldn't help thinking that their pushing back against adversity was key to their survival.

One of our first Michigan chapter efforts, like Dad's in Los Angeles, was to create a local brain bank for Huntington's. Or more accurately, we wanted to add brain tissue from people who had died with Huntington's to the bank for multiple sclerosis brains already established by Wallace Tourtellotte at the university. Brain tissue is extremely valuable for HD research and securing it means persuading people to donate their brains after death. As it happened, I got intimately involved with our first brain donation in Ann Arbor, in a way that I could not have anticipated.

One Labor Day, we got a call from a nursing home in Detroit informing us that a woman with Huntington's had died there and asking if we wanted her brain. "Yes, we do," I told the woman on the line. "Please send the body to us at the university immediately." She replied that an ambulance would drive the body to Ann Arbor right away and told me to stand by to collect the brain when the pathologist removed it. To appear professional and respectful, I decided to put on a white coat over my street clothes, a lucky move as it happened. An hour or so later, the body arrived, but the pathologist we had contacted never appeared. Only his assistant, called a "diener," showed up. After a few minutes, the diener turned to me and said, "We really don't need to wait for the pathologist, Dr. Wexler. I can remove the brain under your supervision." I didn't tell him I wasn't a medical doctor. I didn't tell him anything. I just shrugged and gestured for him to go ahead.

I watched, frozen in a pose of studied nonchalance as he peeled back the hair over the face and sawed away at the head. The sound of the saw cutting through bone made me want to stick my fingers in my ears, but that seemed distinctly un-doctorlike. Finally, wresting the brain from its moorings and placing it in front of me on the laboratory bench, the diener said, "Would you like to take out the requisite tissues, Dr. Wexler?" I studied the brain, scrambling to think of my next move. Not only was this the first time I had ever seen a brain, it was the first time I had ever seen a dead body. "Oh, you go ahead and remove them. I'll put the tissues in the jars," I answered, as though that was how I handled all my brains.

Wearing surgical gloves, I unscrewed the first jar as he handed me a piece of the brain. I put it in and screwed the top on. I could do this, except the men's surgical gloves were too big for me and made the process awkward. It would be much easier without the gloves. Off they came. Just then, the brain started to slide off the table. Instinctively, I saved it from falling with my bare hands. I was instantly seized with a petit mal of panic. What if Huntington's is caused by a virus, as some researchers believe, and I've infected myself? I willed that not to be the case, and we finished collecting the tissues. The samples were sent off to the National Institutes of Health. The virus theory, fortunately for me, was ultimately disproven.

♦ ♦ ♦

The American Psychiatric Association describes implosion therapy as "a technique used in behavior therapy where the client is flooded with experiences believed to be relevant to the client's fear." I would employ a version of this technique to cope with my mother's—and Alice's and my—new reality. I had been struggling to choose a topic for my PhD dissertation, wavering between a topic related to Huntington's and one more in line with my broader interests, such as creativity. In our many letters back and forth, Dad immediately squelched the latter idea. Too vague and impractical, he said. "The urgent thing that occurred to me is that you should learn from Alice and her dissertation troubles." Dad had strong feelings on this subject. "A dissertation is something to get rid of," he wrote. "If you try for something really new, free, creative, you'll be at it for years. It will bore the hell out of you forever, disgust you, and prove of minimal value. You will get turned off from all future creative work ... And anything ... [you write] on the subject of creativity for a thesis committee is bound to be a bitch."[1]

Eventually, I decided on a thesis exploring the experience of living at risk for Huntington's. I called it "Perceptual-Motor, Cognitive, and Emotional Characteristics of Persons at Risk for Huntington's Disease." Martin Mayman, my advisor, knew my story and was very supportive.

He pointed out the counterphobic aspect of my choice and the idea of implosion therapy, which was in the air at the time. He helped me understand that studying the disease was a way to distance myself from it and also to get a handle on it. Other faculty, however, were critical of my topic because it was so un-psychoanalytic—even Dad had reservations, although for different reasons. One faculty member in the department was truly nasty about my choice, saying, "Well, Nancy, just because you study it doesn't mean you're not going to get it."

In the end, though, it turned out to be a good choice because it gave me permission to think deeply about the disease but with a certain distance. I could talk about Huntington's without seeming selfish and whiny. I could talk about how people felt in the context of helping them. I needed this time to sort out my own feelings and I could do so by going around to other people and asking them how having HD in their family affected them. I could mask my fears by asking others about theirs.

I decided to tell my boyfriend, Kenneth, a fellow student in the psych PhD program, about my being at risk. We'd met shortly after I arrived in Ann Arbor. I was attracted to his sensitivity, and he told me he was disarmed by my "openness," a performance that had begun as self-protection and over time became second nature, to the point that I believed it myself.

Kenneth was aware that my father was prominent in the world of psychoanalysis and that, growing up around analysts and an occasional movie star, I was the product of a somewhat exotic upbringing. When we went into Detroit to a Rolling Stones concert, he was amazed to find himself hanging with the band at an after-party in their hotel suite, thanks to introductions from a journalist friend of mine at the *Detroit Free Press.* Although Kenneth was awed and even a bit ill at ease, I felt right at home. My world was more expansive than his, and I loved sharing that part of my life.

Now I was sharing an equally exotic, if less appealing, aspect of my life. The aplomb Kenneth had mustered in suddenly coming face-to-face with Keith Richards deserted him when he was introduced to a Nancy he had never met. When I explained that I was at risk for Huntington's disease and described the symptoms and trajectory of the illness, he was taken aback. Like most people, he knew little about it until this crash course I was providing, so some of his reaction was understandable. But his look of dismay cut me to the core. While he quickly regained his composure to offer support and understanding, his expression in that initial unguarded moment was branded on my brain.

Our relationship continued but it changed, and not in the way my relationship with my mother changed once her "eccentric behavior" was attributed to a disease. After I knew about Huntington's, I looked at Mom with more understanding and empathy. Now Kenneth was looking at me differently too, but not in a way I welcomed. In the middle of the most casual conversation, I would feel I was being watched, as if I were under surveillance for symptoms. I could only imagine the even more distressing ways that an actual diagnosis might alter people's perceptions of me, perhaps leading them to think of me with pity, as a victim, as someone who was diminished, who need not be taken seriously. Candor, I could see, had its costs.

♦ ♦ ♦

I had never imagined that Columbus, Ohio might play an important role in my life. Then one day my father told me about a Centennial Symposium on Huntington's Disease being organized for the spring of 1972, one hundred years after George Huntington gave his historic paper in nearby Middleport, describing what was then called hereditary chorea. A group of neurologists interested in Huntington's from the World Federation of Neurology, along with families touched by HD, including Marjorie Guthrie and my father, were organizing an international meeting in Columbus to assess the current state of knowledge about this disease. To everyone's surprise, nearly one hundred and fifty people showed up, clinicians and basic researchers, coming from as far away as Australia, Japan, Israel, and Norway. Columbus was close to Ann Arbor and Bloomington, and Alice and I both immediately signed up.

What I remember most, aside from astonishment at seeing so many people interested in Huntington's, was a grainy, flickering black and white film showing men and women with obvious chorea, in all stages of the illness, on the unpaved streets of a barrio called San Luis on the outskirts of the Venezuelan city of Maracaibo or poling their *chalanas* around a stilt village called Laguneta in a lagoon of Lake Maracaibo. I was shocked to see all these people with Huntington's, out and about and not locked away at home, though clearly living in conditions of extreme poverty. A young Venezuelan psychiatrist, Ramón Ávila-Girón, a student of the pioneering Dr. Negrette, narrated the film to the audience in English, describing the families in this region of his country where, we would later learn, the prevalence of Huntington's was the highest in the world. As I write these words more than fifty years later, I can still see those powerful, poignant images that first created in me a longing to visit this haunted place, although when and how I could not yet imagine.

♦ ♦ ♦

The ad in the *Detroit Free Press* read: "University of Michigan psychological study seeks people at risk for or diagnosed with Huntington's disease." My research, which formed the basis of my dissertation, aimed "to describe in depth a group of individuals with varied backgrounds across a range of skills with a view toward understanding the variety and pattern of early clinical manifestations, the effects of environment on symptom onset and progression, and the impact of the risk situation itself."[2] I placed the ad with no expectations about the response, hoping to find individuals whom I could interview. Based on my own furtive relationship with the disease, I knew only that *I* would never answer such an inquiry.

Fittingly, the first thing my research revealed was less about the subjects than the researcher. The considerable number of people willing to share their experience with HD placed me face-to-face with my own reticence. Jonathan and Mary Green greeted me at the front door of their tidy suburban home. If they were nervous about meeting me, I was more nervous about meeting them. Jonathan offered his hand. It told a whole story. His firm handshake was fighting a competing energy. Feeling this tension was, for me, an all too familiar experience. It was as if his nervous system had been hacked. As the Greens led me into their living room, I noticed Jonathan had the erratic gait associated with Huntington's. But he was still in charge and fully present.

In designing my study, which used a battery of standard psychological and cognitive tests, I decided that sometime during each interview I would share the story of my mother's recent diagnosis and my adjusting to "at-risk" status. I figured my subjects would be less likely to feel they were simply the objects of medical curiosity if they knew we were in this together, even though I was the clinician and they the subjects. I felt that their knowledge of my status would even the playing field and make the interview more honest and informative. It was not simply about gathering data. I, too, had a stake in the outcome.

We got through introductions, and I told my story. I was relieved to then begin the tests where I could regain my clinical footing. I would conduct these same tests with every respondent, but Jonathan was my first. I started by giving him a series of categories and asking him to name things that fall into that group. "Fruits." And he responded, "Oranges, apples," et cetera. Trees, sports, vehicles. I was looking not so much for the answer but for the process he went through to arrive at it. How he reacted to being questioned: Was he confident, defensive, muddled, clear-minded? Was there a physical impediment to his answering? Was he fighting to mask his disability, or had he given into it? I attempted to maintain neutral expressions and tone of voice as I took copious notes.

I next administered the pen-and-paper part of my test, asking Jonathan to make a line drawing—an abstract drawing, no pressure to replicate reality. He accepted the assignment without reluctance, with a readiness I found to be rare as I met more subjects. In fact, Jonathan's willingness to draw for me was so rare among later subjects that when one of my colleagues asked me to describe how exactly my drawing test worked, I joked, "Not very well."

As I watched Jonathan, I noted that while he was comfortable expressing himself through drawing, his unsteadiness and lack of coordination impaired his ability to do so. I also realized how much I had been hoping for that not to be the case. Jonathan was in his thirties. He may have been my age—twenty-seven—when his incipient Huntington's symptoms appeared. I wanted him to be okay, because I wanted myself to be okay. We're all in this together.

We moved on to the interview. This was a free-form conversation designed to evaluate the effect of HD on my subject's emotional state. The degree to which a person perceives themself as less than well or less than able profoundly affects the physical toll exacted by any disease.

Jonathan was no longer able to work; he was now dependent on his wife to be the breadwinner. I could hear in his voice the toll that disenfranchisement had taken. At the end of the interview, I asked him how it felt to have Huntington's. He took a moment, then answered simply, "It's hard." I wrapped up the test, thanked the Greens for their participation, and left.

Making my way back to my car, I was struck by the stark economy of Jonathan's words: "It's hard." I felt like crying. But I remembered my father discovering me face down on the carpet, sobbing over the death of my mouse. "Nancy, you'll never be a scientist." Instead of weeping, I congratulated myself on proving him wrong.

I've sometimes felt my sensitivity to others as both a gift and a curse. When I approached the next respondent to my newspaper ad, I was determined to emphasize the gift. I knew Patricia Wilder from our CCHD chapter activities. A woman in her forties, she did not have Huntington's, but her mother did. Like me, Patricia was living in the limbo of being at risk. But unlike me, she had a child who could inherit the disease. Caught in a vise of vulnerability between her mother and her child, she had little time to worry about herself. She was also the sole support of her family. With the recent death of her husband, electrocuted in a work accident, she had achieved a trifecta of stress: grief, trauma, and economic insecurity. She recounted all of this matter-of-factly, without emotion—with what some might see as stoicism, but which I saw as flatlining. Patricia's life had left her in shock.

Patricia continued. She told me about her mother, the same age as my mother, institutionalized with end-stage Huntington's. Her mother's two sisters were also affected. The parallels seemed uncanny. I struggled to maintain my clinician's distance. I told myself this wasn't about me. I was simply there to gather data, but that seemed counterintuitive. When I shared the story of my mother's family, Patricia's eyes filled with tears. In the safe presence of a fellow traveler, she allowed herself to feel. I no longer saw empathy as a hindrance to my research. That was Patricia's gift to me.

I went on to meet and develop relationships with a cross section of people affected by Huntington's—from those at risk, like me, to those recently diagnosed, like my mother, to those in the end stages of the disease. Seeing my possible future played out enabled me to see other people's pain and need as a bridge between us, a point of emotional contact. I came to believe that my core purpose as a healer was to make people feel seen. When you're seen, you're no longer alone. When you're not alone, you're not as frightened. And then, the fortune teller has a tougher audience. I was feeling empowered by my implosion therapy and inspired by the collective life force of the people I'd met for my study. I was on the front lines, working to shape my fate.

I found through my doctoral research that none of the at-risk subjects differed significantly in cognitive capacity from so-called normal test controls, even though some were close to the peak age for onset of symptoms. I concluded that "the decline in capabilities may be considerably more gradual than we have appreciated," a thought I found most reassuring.[3] A few years later, I distilled some of my dissertation research into a paper with the title "Genetic 'Russian Roulette': The Experience of Being 'At Risk' for Huntington's Disease," my first professional publication.[4] I was thrilled when Romi Greenson, Dad's colleague, wrote to tell me he had "rarely been so touched by anything I have read in recent years."[5]

Looking back now, I can see how, throughout this study, I was researching and writing my own life as well. When I described how my research subjects recoiled at the idea of being the victim of a random genetic accident, I was describing my feelings. When I focused on how Huntington's shaped the negative self-image of some of my respondents

Genetic "Russian Roulette":
The Experience of Being "At Risk"
for Huntington's Disease

*Nancy Sabin Wexler*

Huntington's disease, (HD) is an hereditary disorder of the central nervous system. It is transmitted through a single autosomal dominant gene with complete penetrance. Symptoms usually appear in adult life between the ages of 35 and 45. However, persons as young as 2 or as old as 80 have been known to develop the disorder. The disease is most often

*An early publication, reprinted by the Commitee to Combat Huntington's Disease (CCHD).*

who had encountered rejection from others, I was conjuring up my own fears too. At the same time, I felt inspired by those respondents who said their at-risk status gave them an enriched perspective on living for the here and now; that it motivated them to focus their energies on activities and relationships that were truly meaningful and gave them the courage to make changes in their lives that brought them pleasure and pride. They were teaching me how to live my life as I was testing them to learn more about theirs.

CHAPTER

# 3

# THE YOUNGEST COMMISSIONER

October 17, 1977. I'm in the Capitol Building in Washington, D.C., at a hearing before the U.S. Senate subcommittee on Labor, Health, Education, and Welfare. Senator Birch Bayh is presiding. Images of people in various stages of Huntington's disease flash across a screen, while Jennifer Jones Simon, a member of the Congressional Commission on the Control of Huntington's Disease and Its Consequences, addresses the room. From the hundreds of hours of testimony we've gathered in hearings held across the country, Jennifer reads one man's statement:

> *How can I adequately convey to you the pressure of living at risk of developing Huntington's disease? Let me start by telling you about two possessions I keep. One is a picture, a photograph from a newspaper clipping. It shows a woman in her 40s. She is strapped into a wheelchair. She is obviously out of control; legs flung high, feet over her head. Her face is a study of anguish, lips stretched, teeth bared, jaws pulled wide, tongue outthrust. Her body is emaciated to the point where every bone and tendon is visible through her clinging skin. She is dressed in a diaper. I keep it as a ready and infallible reminder of what Huntington's disease really has to offer me should it become my lot. My other possession is a .38 caliber pistol. It is my insurance that I will never end up like the woman in the photograph.*[1]

I speak after Jennifer. I introduce myself as someone also living at risk—this blonde, blue-eyed California girl, bursting with vitality—informing them, "I could have written that man's words." I proceed to read my diary entry describing my mother lashed to her chair with sheepskin padding tied to her forehead. I hold up an enlarged photograph of another woman in that same state.

*A rare moment of levity during our testimony before Congress. (*Left to right*) Me, Jennifer Jones Simon, and Guy McKhann, neurologist from Johns Hopkins University.*

"I share that man's fears," I tell the senators. "I'm haunted by the same images. Like him, my mother's agony possibly awaits me. But unlike that man, my contingency plan is not a gun to my head. It is finding a cure for this deadly disease. When my fears come up—and they do—I turn my attention to that day's work. And my work for today is to come here with my friend and implore you to support our cause."

♦ ♦ ♦

Three years earlier, I was living in a studio apartment on West 9th Street and 6th Avenue in the heart of New York's Greenwich Village, looking right into the historic Jefferson Market Library with its glorious clock tower and Gothic pinnacles and parapets. Below me, at street level, Balducci's, a fancy deli, catered to the best-fed cockroaches in the city. But I didn't care: I was twenty-nine and an assistant professor of clinical psychology at the prestigious New School for Social Research, an institution with an illustrious lineage. Founded as a graduate university by refugees from Nazism in the 1930s, the New School boasted a faculty of distinguished scholars, and I was excited to be among them. Right away I began putting my research on Huntington's out into the world. I was thrilled to give my first paper at the 1975 annual meeting of the

American Psychological Association in Chicago with Dad cheering me on from the audience.[2] I also continued to work with CCHD and travel to Los Angeles to attend foundation workshops.

Nonetheless I did not share my at-risk status nor discuss my advocacy work with my New School colleagues. I had even asked my advisors at University of Michigan not to mention my at-risk status in their letters of recommendation (which would have been inappropriate anyway). My interviews with a wide range of people at risk for and living with Huntington's had taught me how a genetic condition such as HD can shape the ways people see and react to you and how they treat you. I had learned that there were good reasons why members of HD families tried to hide the condition, and I also learned the high emotional costs of keeping it hidden. As I grew increasingly busy with teaching and setting up a clinical program at the New School, I could feel myself wandering away from the Huntington's path, and that felt good.

♦ ♦ ♦

One day I received a phone call that changed the direction of my life. A neurologist named Tom Chase was inviting me to join a new congressional commission on Huntington's disease as deputy director. Tom, with whom I would later become good friends, was then scientific director of the Intramural (in-house) research program at the National Institute for Neurological Diseases and Stroke, or NINDS (formerly National Institute for Neurological Communicative Diseases and Stroke), known informally as the neurology institute, part of the National Institutes of Health. He had strongly supported the creation of an HD advisory commission. There were many such commissions created by Congress to study problems, including diseases, and make policy recommendations. Marjorie, my father, and a group of HD families from New Jersey had pushed hard for a Huntington's commission; already they had helped move HD from an obscure disease no one had heard of to one that Congress considered worthy of funding.

Tom's invitation sounded enticing, but it created a lot of anxiety in me. I had just started getting to know my colleagues, just begun enjoying the wealth of theater, music, dance, and museums that New York City had to offer. Did I really want to leave my new life here, with all its freedom, intellectual excitement, and cultural offerings? Mostly, though, I worried that I could screw up the job. A successful commission could help bring new resources to Huntington's research and care. What if I ruined it? I remember one day when I was visiting Mom in California, I went to the beach and walked in the surf in my rolled-up jeans, weighing all the dimensions of this choice. My thesis was done. I was starting a career in clinical psychology, totally away from anything having to do with Huntington's if I chose, and I welcomed that escape. I still felt uncomfortable talking about HD. I could decline the invitation and everyone would understand. Even Dad said, "If you want to do it, fine. But if you don't, that's also fine. Don't do it out of guilt. It's more important to lead your life." So, I could get out, I could Not Do This and people would understand.

But then I saw the faces of all the Wilder kids calling me. They believed in me and in the mission to find a cure. How could I turn my back on them? A commission could really have an impact. Also, the fact that prominent neurologists at the neurology institute, like Tom, knew I was at risk and accepted my status gave me confidence. He wanted to work with me; others at the neurology institute were willing to take a chance on me. I reasoned that as deputy director I would at least be working under someone more experienced. And Marjorie as chair and Dad as co-chair would be there to provide additional support. Eventually, I said yes.

I didn't immediately take a leave of absence from the New School, so again I became a commuter, this time by train between New York and Washington, D.C., once more toggling between two worlds. We were just beginning to organize the commission's membership when I received another life-changing phone call. My father was on the other end of the line. He explained to me that the commission's executive director had just told him he couldn't take the pressure; if forced to continue, he would commit suicide. Now Dad was changing his tune. "Nancy," he said, "We need you to take over." At barely thirty years old, I felt like I was in one of those movies where the pilot has a heart attack and the passenger is asked to fly the plane.

Over the course of the next few weeks and after conversations with many people, I decided that I did want to take on this new challenge, with all the opportunities it offered for learning and for making a difference. New York could wait. What I lacked in experience, I would make up for in passion. Once I decided, I never looked back.

♦ ♦ ♦

Along with Tom Chase and Don Tower, the NINDS director at the time, I began compiling a list of possible commissioners. Right away, we had a fight with members of the conservative Gerald Ford administration over who could join the commission. Administration officials had initially vetoed Marjorie because, they said, Woody Guthrie had been a communist! We insisted she must be on the commission; in fact, she should be chair. Eventually they agreed to accept her as chair along with my father as the co-chair. We also had a fight over Guy McKhann, who was the Kennedy Professor of Neurology and chair of the neurology department at Johns Hopkins University School of Medicine; they did not like any association with the liberal former president John Kennedy! But McKhann stayed. The other HD family member, apart from Marjorie, Dad, and me, was Alice Pratt, the wealthy Texan who had started her own Huntington's chorea foundation after her husband developed the disease. Dad always liked Alice on account of her Texas iconoclasm and how she drove around the state talking on her CB radio with truck drivers who knew her as Cricket. I liked her too.

Overall, the commissioners were a formidable—and friendly—group, mostly twice my age, the majority men, of course. They included a medical school dean, two chairs of neurology departments, and the director of a so-called "handicapped" unit at the Minnesota Department of Health.[3] Later on, Jennifer Simon joined the commission. Dad thought that the commission might help distract Jennifer from her grief over her daughter's recent death by suicide, however self-serving that argument may seem. Jennifer understood the democracy of depression and mental illness, and she knew that fame and money offered no protection.

Right from the start, we decided to approach Huntington's as a prototype for other major neurological, psychiatric, and genetic disorders. We wanted to emphasize that understanding Huntington's would not

*Members of the Commission:* (Left to right, clockwise) *Dad, Ching Chun Li, Stanley M. Aronson, Charles MacKay, me, Guy McKhann, Jennifer Jones Simon, Stanley Stellar, Lee M. Schacht.* (Center above) *Marjorie Guthrie;* (center below) *Alice Pratt.*

only give insights into the nervous system and mechanisms of inheritance but also would illuminate problems common to other neurodegenerative and psychiatric disorders. We would show that people with Huntington's were part of a large community in despair over the inadequacies of our fragmented health-care system. We organized a series of working groups, which fell largely into three areas: basic research, the role of drug companies in developing therapies, and insurance coverage for long-term care.

One of the basic research groups, focused on genetic linkage, turned out to be much more central than we anticipated. Linkage refers to the tendency of genes located physically near one another on a chromosome to be passed down together from one generation to the next during the formation of the sex cells—egg and sperm. During that process, called meiosis, pairs of chromosomes separate and pieces of the maternal and paternal chromosomes may exchange places (crossing over), so that everyone ends up with a combination of genes inherited from both parents rather than having an exact replica of the genes from one parent or the other, a process called recombination. This reshuffling of genes ensures that there's a little bit of variation in each human being. And genes located far from each other on different chromosomes are especially likely to get reshuffled. However, when genes are located near each other on the same chromosome, they tend to remain together and get passed down together, and then they are said to be linked.

*Executive Director.*

I used to explain linkage and recombination, taking some creative liberties, in terms of penguins hanging out together on the ice in the Antarctic. If two penguins are standing far apart on an ice floe when it breaks into pieces, the penguins could very well become separated. Then when the pieces eventually come back together again to form new ice floes, the separated penguins can end up on different floes among new neighbors—a recombination event. But if they are close together on an ice floe, they are likely to remain together—linked—when the split-up floes rejoin each once more.

Identifying linked genes and calculating the frequency of recombination events between two genes—determining how often two genes get separated in multiple meioses—were two of the early methods geneticists developed for mapping genes on chromosomes, initially the chromosomes of *Drosophila,* or fruit flies, using eye color or wing length as visible markers for the genes that they could not see. But by the 1970s some geneticists were using linkage to try to map human disease genes. If, for example, everyone with Huntington's in a large family had type B blood over several generations and everyone who did *not* have Huntington's had type A, O, or ABO, we could infer that the gene for HD and the gene for blood type were on the same chromosome, and near each other; they were linked. Blood type could serve as a proxy, or genetic marker, for the nearby presence or absence of the disease-causing gene. It could be used to predict, in advance of any symptoms, who among those with an affected parent was destined to develop the disease.

Geneticists had long been interested in identifying such a marker for its eugenic value: to discourage or prevent those carrying the aberrant version of the gene from passing it on. However, finding a means of prediction had broad support among families with HD as well. Many people at risk for Huntington's wanted to learn their genetic status, to plan for the future, making childbearing decisions, or simply to escape the anxiety associated with the uncertainty of a 50–50 risk. They were also interested in a prenatal test.

P. Michael Conneally, a geneticist at the Indiana University School of Medicine, had been trying to find a marker linked with Huntington's even before the commission began. One of Mike's graduate students had gotten him interested and for several years they had been searching using blood samples from Huntington's families in the United States, especially one large family from Iowa whose members had been exceptionally generous with blood donations. Looking among the traditional biochemical markers such as blood type and certain proteins on the surface of cells, Mike and his students had figured out where in the genome the gene causing Huntington's was probably *not* located, useful information in itself. But they were running out of markers to test. They were also running out of large enough families who had inherited the gene from a common ancestor—necessary for these gene linkage studies because descendants of a different ancestor might have a distinct form of the marker. Mike and his students were coming up against a brick wall.

♦ ♦ ♦

A second working group addressed presymptomatic testing directly, a topic which not surprisingly turned somewhat contentious. This group had a broad mandate: Working Group on Molecular Genetics, Immunology, Virology, and Presymptomatic Detection, although the last in this list threatened to subsume all the others. In fact, the topic of predicting future Huntington's, prenatally and in people who currently had no symptoms, loomed over the entire commission and threaded through several of the working group reports.

In their final report, this group assigned "highest priority" to developing a test and predicted that a large majority of those at risk for Huntington's would use such a test were it to become available. Yet most of the working group members took a surprisingly cautious and measured approach to the topic. They developed preliminary guidelines for the use of a test once it was validated, emphasizing that testing should be entirely voluntary and that no one should be tested without their consent. Genetic counseling should be nondirective, aimed at helping those considering testing make decisions that were best for their individual lives. The group's conclusion bears repeating: "If a presymptomatic test is developed and validated in the absence of effective treatment for Huntington's disease, the test results have the potential for disastrous consequences as well as good."

Just one member of the working group approached the test from a straightforwardly eugenic perspective, framing the predictive test as a means of disease prevention rather than as an option for the benefit of the person at risk. In this view, the test could be used to identify those carrying the aberrant gene so that they could be discouraged or prevented from passing it on. This argument reflected the differing perspectives of public health and clinical care while highlighting just how persistent eugenic thinking remained even in the late 1970s.

♦ ♦ ♦

Although it was not an initiative of the commission, the commissioners and I strongly supported and helped to shape the National Genetic Diseases Act of 1976. This legislation mandated funding from Congress for programs to train genetic counselors, social and behavioral scientists, and other health professionals. Genetic counseling had typically been performed by geneticists, pediatricians, or obstetricians, who saw their role narrowly as imparting genetic information as accurately as possible, usually in the context of a pregnancy. They weren't trained in understanding the emotional nuances of such information nor in navigating the complicated psychological dimensions of imparting such information. They weren't trained in the psychological and social skills we felt were essential to the role of a genetic counselor. We argued that genetic counseling was not merely a matter of communicating scientific facts. It was also about helping clients understand them, both intellectually and emotionally, and integrate the information into their lives to make informed choices. The first graduate genetic counseling program to address the psychological as well as the scientific dimensions of counseling had been established only in 1969, at Sarah Lawrence College, although soon there would be a number of such programs at universities around the country. To assure the highest level of service, we urged that all genetic counselors be required to hold a Master's degree and work in collaboration with a medical doctor. I felt gratified that my research at Michigan had helped prepare me for a role in developing these programs.

♦ ♦ ♦

Of all the commission's working groups, one was especially close to my heart. Ever since the Centennial Symposium in Ohio five years earlier, I had been haunted by the images we had seen. I had felt then, without articulating it, that those Venezuelans would be a part of my future. Now the Venezuela Working Group was officially acknowledging their importance for science. The group's report urged that Congress appropriate funds to the NINDS "to collaborate with Venezuelan scientists to design, support, and conduct an interdisciplinary study of the population affected by Huntington's disease in the state of Zulia, Venezuela."[4]

It mattered a great deal to me that I wasn't alone in wanting to travel to Laguneta, Barranquitas, and San Luis—the three communities with the greatest number of families with Huntington's—to see and meet the people for myself. Now, high-level members of a U.S. Congressional commission were ranking such a study "of highest priority" and stating that the results of this project "could radically alter the direction of Huntington's disease research for investigators world-wide."[5] But would the people of Laguneta be willing to participate in our research, to share their knowledge of their family histories, and donate their blood and skin for science? How would the residents of San Luis feel about being subjects of filmed neurological exams? What aid and assistance could we offer them in exchange for what we were asking of them? And what if Huntington's in Venezuela was different from the disease everywhere else?

I knew even before the commission ended that I would travel to Venezuela to find answers to these questions. But I didn't know that, a year and a half later in July 1979, I would make that exploratory trip to Lake Maracaibo with Tom Chase to visit these communities. Nor could I have envisioned that for the next twenty-two years I would spend six to eight weeks each year in Venezuela, returning home reluctantly each time to the world of government and academia where I could be an instrument of change from that direction. And never could I have imagined that every time I went back to Lake Maracaibo, it would feel like coming home.

♦ ♦ ♦

Along with setting up the working groups of experts, we decided early on to hold a series of public hearings around the country—eleven in all—where HD family members could testify, as could those who dealt with Huntington's in their professional capacity, including social workers, neurologists, insurance representatives, police, teachers, nurses, and public officials of all kinds. The hearings often took on the emotional aura of a revival meeting as HD family members who thought they were the only ones in the world with this disease met others and learned they

*Dad and me during the commission.*

were not alone. Their words formed the basis of our recommendations and the heart of our testimony before the senators at the conclusion of our task. All the issues we heard during the hearings helped us form speaking points when, finally, we began addressing the Senate panel.

Certain themes came up over and over in the testimonies, mostly painful and all too familiar. I had experienced almost all of them: the stigma and silence surrounding Huntington's; the ignorance of most doctors about the disease, including about the hereditary pattern; the denial on the part of family members who knew about the disease but refused to discuss it; the fear and anxiety suffered by those at risk waiting for years to see if symptoms would emerge; the anxiety that should a predictive test be developed, it could reveal future onset but do nothing to forestall or prevent it; the divisions within families over how to respond to the illness; the sense of belonging to a tainted lineage; the shame and embarrassment about relatives with the symptoms; the tremendous financial burden of care; and the sense that there was nowhere to go for help. From the testimony:

> *My grandmother, mother, uncles, sister, and brother all have Huntington's, but only under the most tense situations do we ever discuss it. It's never discussed with the patient.*

> *My son was fourteen years old when we were told his father had HD. I had never heard of the disease before that. That same evening his father took gopher poisoning, but it was too slow and too painful; he was found leaning on the horn of his car with all the doors locked. After that he was arrested for drunk driving (it was HD as he didn't drink), could not hold a job, pawned everything he owned, and when that ran out he started writing bad checks. He signed himself into the Norwalk state hospital, but that was of no help to him, so he signed himself out, rented a motel room, and shot himself in the head. For twelve years now, I have lived in fear that our son will get the disease.*

To the senators, I pointed out how Huntington's often robs people of their ability to work in the prime of their lives, undermining their sense of purpose and dignity. I told them about Jonathan Green, the man with Huntington's from my Michigan study, and the loss of employment that cost him his sense of purpose and self-respect. Along with his health, he'd lost his place in the world. His life had been reduced to just two words: "It's hard." Another case in point: "He's dependent on others, mainly his immediate family. He's bored stiff most of the time; he cries a lot. He feels he has nothing left to live for. He's only 36!"

I related the stories we heard often about people being misinformed about heredity, with some reporting that they were advised the disease was transmittable only by males and others told that only women get Huntington's. And sometimes people were denied information altogether: "The doctor told my mother when her mother died, that it was none of her business what she died of."

I told the Senators how we heard repeatedly of young people being advised by their doctors against having children and the traumatic impact of that advice: "It's every girl's dream to have a baby, and all of a sudden I was advised not to have a child." And at another hearing: "It was at his death that my brothers, sister, and I were told about my father's hereditary disease and of course, warned not to have children. The warning came years too late...". The advice to get sterilized surfaced as well, echoing what some physicians were writing in the medical literature: "There is probably no other disorder with such a strong argument against reproduction." We heard this not only from physicians but occasionally even from family members themselves. "My wife's relatives still only mention HD in hushed tones and only to other family members. It's time to get HD out into the light! If potential victims have seen HD in their family, the way we did, getting sterilized would be a cause for rejoicing!"

The severe financial burden of Huntington's for families and the need for paid, long-term in-home care, I emphasized, was a necessity that extended far beyond Huntington's disease. I told the senators how, over and over, families unraveled once the financial thread was pulled. Medicare might pay for a limited number of days in a nursing home following a hospitalization, and Medicaid also covered some types of nursing home care, depending on the state. But to qualify for Medicaid, the patient had to possess minimal assets and have an income near or at the poverty line. The avalanche of expenses during the ten- to twenty-year course of Huntington's, I explained, threatens to bury everyone, save perhaps the uber-rich.

*I would either have to take him home or I would have to place him in a state institution. If I wanted a private facility, I would have to liquidate every bit* *of finances that we had. Every bit of income up to $2500. And I would be allowed only $300 a month to live on. Therefore, I would have to quit my job and could no longer earn my salary because it would be over and above the required amount. I could not afford to place my husband into the private facility or even the state facility because I couldn't afford $1200 a month. I still had, at the time, a daughter at home, that was going to school and I had all the other personal needs of any human being who is alive and well and trying to live as best they can.*

I related the story of my own family: how the diagnosis of my mother's brothers had put the onus of responsibility on my father and impacted my parents' marriage. And that was just the beginning. Anxiety over the ability to pay long term for my mother's care as the disease progressed threatened to supersede concern for the care itself. And our family was well off. At one point, after my parents' divorce, my father considered asking her to remarry, thinking it might ameliorate her situation. But he quickly jettisoned that idea when he was advised that remarrying her would jeopardize her ability to receive state financial aid should the need arise. And so, a crucial decision about my mother's quality of life was decided by the balance sheet. As one frustrated family member put it, "There is something wrong with a society that says that a person with a debilitating disease, for which there is no known cure, must not only carry the burden of his fate, but must also be financially penalized for having that illness."

Too often, I reiterated, the testimony of family caregivers was a simple statement of their predicament, as if they dared not hope for a way out. I wanted the Senate panel to understand the thoughts and fears of a woman who testified at one of our hearings: "Sometimes I'm afraid that my husband will commit suicide. And sometimes I'm afraid that he won't."

♦ ♦ ♦

By the time the commission was concluding, my anxiety over my responsibilities as executive director had abated. I'd come to enjoy the position and the political leaders I was meeting—for instance, Senator Birch Bayh of Indiana and First Lady Rosalynn Carter, always a strong advocate for mental health. Our presentation in the Senate and Marjorie's in the White House evoked much enthusiasm from the senators and their aides and assistants and also from the press. I felt exhilarated after the long day addressing the Senate panel. It helped, of course, to have a glamorous movie star in our group. I'm certain Jennifer opened doors for us that might otherwise have remained closed. She was a formidable and passionate advocate. Together we were quite a pair, I having inherited my father's talent as a snake charmer and learned his skills at persuasion and Jennifer being, well, Jennifer. My friend Elaine May expressed it succinctly: "You just wanted to help her."

Because all this took place in the 1970s, the senators we faced were predominantly older men. But my parents had raised my sister and me to demand gender equality, and that attitude served me well in Washington. I never went in looking for a fight, but I was prepared to make myself heard. And Jennifer—having survived and triumphed in old Hollywood where the prevailing policy was "actresses should be seen but not heard"—had also learned how to speak up and speak out. We had each developed our own version of a charm offensive for disarming potential resistance. We never showed our disgust at being patronized as "you two little gals"; we just made our case. Nothing could deter us from our strategy of delivering an informed message from the heart.

♦ ♦ ♦

As the commission wrapped up its work early in 1978 and we finished drafting our lengthy conclusions, we came to a difficult realization. We had heard about commission reports with complex analyses and hundreds of recommendations that ended up on the shelves of the neurology institute where no one paid attention to them. We wanted to keep our report simple, with a small number of recommendations that had a chance of being enacted rather than a large number that likely did not.

And so we sat down and rewrote the first (overview) volume (not the working group reports), highlighting those of our findings that were relevant to *all* neurogenetic diseases. We wanted our report to serve as a template for both government programs and private initiatives, as well as be a record from the 1970s of the advice given by a group of experts. We highlighted the concept of Centers Without Walls to combine HD research with clinical care. Eventually two were

*At the National Institute of Neurological Disorders and Stroke (NINDS).*

established, one at the Johns Hopkins School of Medicine and the other at Harvard Medical School, each of which has played a major role in advancing both care and research. We called for a Huntington's research roster, which Mike Conneally helped establish at Indiana University in Indianapolis in 1979. The roster allowed HD family members to donate biological samples and give clinical information that would then be preserved and made available to researchers. To address the need for places where patients could go for social interaction and activity while giving caregivers time off, the commission called for the establishment of respite centers. We asked for funding for in-home care to keep patients out of nursing homes. We further recommended legislation and federal funding to establish more graduate training programs in genetic counseling. And we urged funding for small grants, conferences, and symposia on Huntington's, specifically, and on genetic and neurological disorders more broadly. In the end, the commission's recommendations were extremely practical, very specific. We felt that while they focused on this one disease, they were applicable to many more.

♦ ♦ ♦

Despite our rewriting, it became obvious that nothing would be done about the commission's recommendations unless someone at the neurology institute was dedicated to carrying them out. So once the report was completed, my friend, neurobiologist Katherine Bick, known as Kit and by that time a deputy director at the NINDS, asked if I would stay on in Washington as a health sciences administrator at the institute. Apparently, hiring me met with some resistance as I was known as a free spirit, but nonetheless I was hired, and together, over the next few years, we succeeded in implementing many of the commission's recommendations, at least those that were specific to Huntington's disease.

Proud of the commission's successes, I took the published commission report to Los Angeles in the spring of 1978 to show it to Mom. We

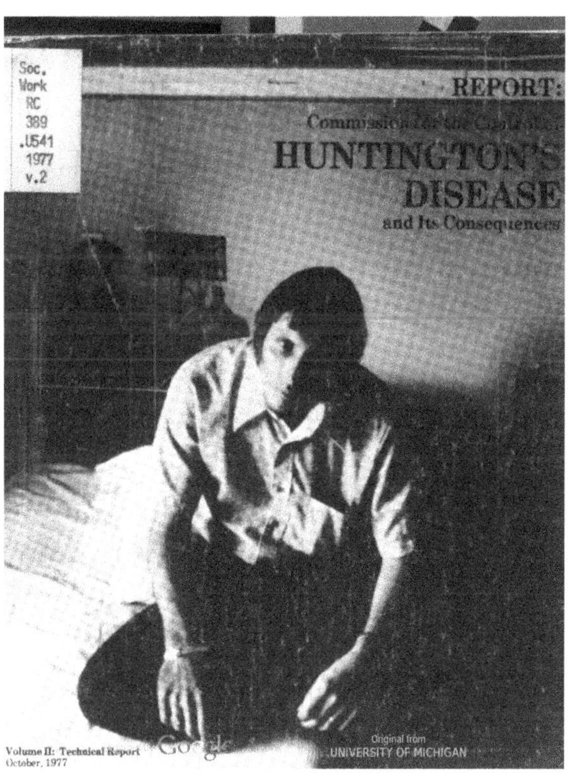


*Volume II of the Commission report.*

had dedicated it to Woody (Marjorie's husband), Fletcher (Alice Pratt's husband), and my mother Leonore. Mom was now completely bedridden in the late stage of the disease, her movements calmed at last. She could no longer speak. But she smiled. She understood. I hugged her for a long time.

*Sunday, May 14, Mother's Day, 1978*

*My mother has died tonight. She was so ephemeral and yet so real in life—a tremendous force which has guided and impinged and constricted and shaped my life has left—and I miss her.*

CHAPTER

4

# LEARNING FROM LAGUNETA

*October 1979, Bethesda, Maryland*

I am sitting at a Hereditary Disease Foundation workshop as the discussion devolves into a yelling match, with people standing up at the blackboard scribbling furiously, arguing their opposing theories. The freewheeling, no-holds-barred style first encouraged by my father is in full play.

David Housman is defending the idea of using the newly discovered RFLPs (pronounced *riflips*)—restriction fragment length polymorphisms—as markers to search for linkage with the Huntington's disease gene. The foundation's new scientific director, Allan Tobin, a UCLA assistant professor of biology, has invited Housman, his geneticist friend from MIT, to explain how the new molecular techniques Housman and his graduate student James Gusella had been developing might be relevant to Huntington's research.

Housman explains that bacteria produce restriction enzymes, which are proteins that serve as a defense mechanism against foreign DNA. They "recognize" foreign DNA and cut it at specific sequences, rendering it harmless to the bacteria. He, along with some other forward-looking geneticists, have figured out that, because these enzymes also cut human DNA at specific sequences, they can be used to reveal places in our DNA that vary between individuals, much as eye color and blood type vary. They propose that these variable segments of DNA, known as restriction fragment length polymorphisms, or RFLPs, can be used as markers in linkage studies to map disease genes.

The method involves taking a short single strand sequence of DNA called a probe (from a "library" of probes made up of unique fragments of human chromosomes) and seeing how it binds to an individual's DNA that has been "digested" by a restriction enzyme. If the probe shows variation between individuals in the size of the fragments, that variation is

*David Housman—"Toughest PI on the floor." (Photo by Alice Wexler.)*

likely due to a polymorphism at a restriction site bound by the probe, an RFLP. And if that's the case, the next question is, do the different sized fragments revealed by the probe track with the presence or absence of a disease in a given family over several generations.

David Botstein, a colleague of Housman's at MIT, thinks it's too great a long shot to search now for a marker linked to one specific gene that could be almost anywhere in the genome, although he, too, is keen on the RFLP approach for mapping. He and several colleagues have a paper in the works proposing to make a map of many markers scattered throughout the human genome that could be used as signposts to map individual genes.[1] They propose that identifying approximately 150 polymorphic markers as landmarks throughout the genome should make it possible to use linkage to make a map of all human genes. Essentially, they want to take a step back. They say that it makes more sense to identify these polymorphisms—these potential markers—*before* you start looking for linkage to individual genes. Their arguments are compelling: To look for a marker for Huntington's that could be anywhere in the genome is like searching for a single person in the United States without a street address, or even the name of their street, city, county, or state. What are the chances of success?

Botstein considers Housman's idea of searching for the Huntington's disease gene premature, a wild goose chase. He scoffs that it would be irresponsible to set up unrealistic hopes in families donating their DNA to the research. It's a question of timing. Housman's approach will eventually work; no one questions that. But most of the workshop participants believe it will take ten years or more. Better to wait for a rough map of the entire genome before homing in on one specific gene.

Mike Conneally, also at the workshop, recognizes immediately that RFLPs constitute a nearly unlimited supply of markers and a huge new resource for mapping genes. Having failed to find linkage to Huntington's with any of the traditional markers, he is eager to try *riflips* for finding this gene. He supports Housman's strategy with enthusiasm. As with the traditional markers, though, everything depends on our having access to large enough families with the disease.

As for me, I fall instantly in love with Housman's bold approach. Even if it takes ten years or more, better to pursue now what seems like a winning strategy than sit around waiting for a rough map of the genome that also might take years. Who knows, maybe we will get lucky! And there is something else: In the midst of everything, I have an epiphany. If Housman's linkage study depends on having access to large families, then maybe I hold the magic key. I'm thinking of the Venezuelan families with *el mal* whom I had met just three months earlier, in San Luis, in Barranquitas, and in Laguneta. I'd spoken with them, hugged them, heard their worries and concerns for themselves and especially for their children and grandchildren whose lives hang in the balance. And perhaps I can contribute to the mapping project, not only as a supporter but as an active participant in the research.

There have been few times in my life when I have felt convinced something was absolutely right, when my heart raced and leapt into my throat, times when I couldn't sit still and wanted to run around as fast as I could, laugh wildly, or explode. I have this feeling at the end of the workshop. I ask David Housman: "In terms of gathering samples for DNA, is bigger really better?" Without missing a beat, he answers, "Definitely. The biggest family you can possibly find."[2]

Suddenly everything has changed. Three months earlier on our exploratory trip to Venezuela, Tom Chase and I were focused on finding someone with a double dose of the aberrant gene. Now a whole new approach is opening before us. Without waiting to hear if the NIH will fund Housman's proposal (submitted by Harvard Medical School Chair of Neurology Joseph Martin as part of a Harvard Center Without Walls grant), the Hereditary Disease Foundation votes in January 1980 to provide seed money to start the project right away.

♦ ♦ ♦

By the spring of 1980, I had bought a small row house in the Mount Pleasant neighborhood in Washington, D.C. I felt gratified at our progress in implementing many of the HD commission's recommendations. And living in Washington had other benefits. It was during a meeting at the NINDS that I met Edward Kravitz, who helped demonstrate the role of GABA as a neurotransmitter in the early 1960s and was a key figure in the department of neurobiology at Harvard Medical School. Ed would have a great influence in the foundation as a science board member and in my own life as a friend. After hearing Marjorie Guthrie and me speak at the meeting about the need for more disease-related science, Ed took our message to heart and invited us to Harvard to discuss how to interest young investigators in studying diseases. As Ed wrote in his memoir, our presentations made him recognize that he and his colleagues hardly ever mentioned neurological diseases in their courses. "As far as our students were concerned," he wrote, "As far as we were teaching them, the nervous system always functioned properly." Our discussions eventually led Ed and me and several others to develop a two-day Neurobiology of Disease Workshop incorporating some of the time-tested features of Hereditary Disease Foundation workshops. These included patient presentations, lectures that were *not* research seminars, and freewheeling discussions. We offered the course at the next annual meeting of the Society for Neuroscience, in 1981. It was so successful that we managed to keep it going for the next forty years. Ed used this experience to create a similar course at Harvard Medical School, one that got the highest ratings from the students of all the graduate courses offered.[3]

I was thrilled that hundreds if not thousands of young neuroscientists were getting exposed to the neurobiology of disease through our course and might decide to focus their research in that direction and, also, that the philosophy of the Hereditary Disease Foundation was spreading to courses at Harvard and beyond. But working as a health sciences administrator at NINDS, I found myself doing tasks that felt less and less creative. I was dealing with grants and grantees, going on site visits, advising applicants about how to improve their grant proposals, and helping to shepherd some of them through the grueling NIH application process. I felt I was effective as a catalyst for the accomplishments of others, but I was feeling depleted, sometimes even cannibalized, holding too many things together and leaving little for myself. I was not feeling challenged in ways I found nourishing. I was glad that I had come and felt I had learned a tremendous amount and had made things happen for HD research that wouldn't have happened otherwise. I had made cherished friends. Nonetheless, I dreamed about script writing, science writing, and even, for a time, going to medical school. "I want the zeal of a scientist or an artist in what I do," I told Dad. I was feeling that zeal less and less.

Returning to Venezuela for the gene mapping project was a high priority, no matter where I lived or what else I was doing. But it wasn't a solution to my life. Dad and I corresponded often about what the possibilities might be. He was a wonderful interlocutor—patient, encouraging, and loving but also frank and direct—suggesting that perhaps I'd had too much of Huntington's disease and should take a break and do something else for a while.[4]

One fear lurked at the edge of my consciousness. "I want to be passionate, dedicated," I wrote my father. But sometimes "you and I seem so close in our interests that I think it frightens me a bit. Even in our talents, which makes our common interests more legitimate... But when I was at Michigan training for shrinkdom, I used to sometimes wonder where you stopped and I started."[5]

Dad understood. He didn't lecture me or deny my anxieties. He also didn't mince words. "It seems blatantly empty and destructive to continue where you are," he told me, "Or up or down the ladder at NINCDS. I can see the pleasure in power, visibility, geographical mobility, meetings, people, contacts, etc."[6] But, in his view, such a life lacked the "growth and creativity that comes from a steady involvement in something beyond the midwifery of middle management." For him, an administrative job in Washington was "the end of the line." Dad's preference, he wrote, "And this is a 35-year-old me talking—I think you should give thought to some creative enterprise. Writing—producing—photography—raising cattle—ranching—making huge amounts of money (with dedication it can be done), etc., etc." My problem, he suggested, was that I could not let go of anything, not of people, things, or activities. Not out of obsession, reservations, or fear of decision but rather because I could not bear to let go of enjoyment, pleasure in whatever I was doing at the moment. I became so overloaded and overcommitted that my life turned into chaos. "Cut the weeds," he told me, "The scrub oak, the poison ivy, the overgrown vines. For you—for sure—less is more!"[7]

I didn't share Dad's grim view of Washington or his sense that working in government could not be creative and rewarding. But once the commission ended, my job became more amorphous and less satisfying. I was definitely looking for something else. Together we weighed the options: staying at the neurology institute; coming to Los Angeles, possibly to run the Hereditary Disease Foundation in his place, although that would mean more immersion in Huntington's; returning to my teaching position in New York where the atmosphere was better ("You were more of everything in N.Y.," Dad told me); or pursuing some sort of creative endeavor anywhere.

♦ ♦ ♦

Over the course of the next year, despite my reservations, I decided to leave the neurology institute and return to Los Angeles to serve as vice president or executive director of the Hereditary Disease Foundation. The new Jennifer Jones Simon Foundation offered me a leadership position as well. But an unexpected invitation changed that trajectory. Perhaps it was the crisis of identity I was going through that persuaded Dad to invite me on a study trip, organized by the Simon Foundation, to visit China's psychiatric hospitals in November of 1981. He would be going on the trip along with other mental health professionals and Jennifer as well.

The China trip turned out to be another turning point in my life, for it was on this trip that I met Herbert Pardes, a psychiatrist from New Jersey, who was at that time the director of the National Institute of Mental Health (NIMH). Outgoing, thoughtful, and enthusiastic, Herb was one of the smartest men I'd ever met. He was also ten years older than me and married, although he was leaving the marriage, or so he said. Dad always advised me, "Tell 'em to get lost until they get unmarried." This time, though, I ignored his advice. Herb had many traits I valued greatly: Besides his generosity of spirit, his compassion, and his terrific sense of humor born of growing up in the Catskills hotels his parents had managed, he was already a father. He had three handsome sons. Soon I would see from his interactions with them what a great father he was. I, too, passionately wanted kids, although ever since Mom's diagnosis, I had put that desire on hold. But if and when we found a cure, or a means of preventing HD, or a test that guaranteed I wouldn't pass it on, then I would have the perfect father for my kids of the future.

These considerations weighed heavily in my mind. The idea that any partner of mine might be deprived of having children and *also* have to deal with a Nancy with Huntington's was unbearable to me. I felt it was unfair to subject anyone to such risks, even if they were willing. Alice used to say to me, "Why don't you let *them* decide? Why decide *for* them? What if they feel you are worth the risk?" Still, I couldn't let go of the feeling that I would be laying too great a burden on my partner. But because Herb already had children, I wouldn't be depriving him of fatherhood in case my dream of having kids did not come true. He was also a doctor, a psychiatrist. He knew what he was getting himself

*With Herb Pardes in China, November 1981.*

into, and he was willing to take it on. I somehow knew he would never abandon me in the way that many spouses—usually husbands—abandoned their partners who developed Huntington's. We had heard many such abandonment stories during the commission hearings. Of course, Herb assumed that I would not get Huntington's; at least, he clung to his denial for as long as he—we—could. Dad too periodically stated that he believed Alice and I wouldn't get it, although he was always making contingency plans *just in case*. Sometimes I believed his optimistic prognoses, and sometimes not. I, too, was a great practitioner of denial.

After meeting Herb, I decided to stay in Washington at the neurology institute, at least for the moment. I could still organize visits to Venezuela. I could still work with the Hereditary Disease Foundation long distance. Besides, I traveled often to see Dad and attend workshops in Los Angeles where I could still play a role at the foundation. One day when I returned to Washington from attending a conference or from Venezuela or a trip to see Dad—I no longer recall which it was—I found that Herb had moved into my house on Monroe Street. He never left. In 1984 we would move together to New York City, Herb to accept a position as chair of the psychiatry department at the College of Physicians and Surgeons of Columbia University, and me to take a joint position in the departments of neurology and psychiatry at the same institution. We would move into an apartment at 15 Claremont Avenue—just a few blocks from Teachers' College of Columbia, Dad's alma mater, and from Morgan's fly room in the zoology department where Mom had studied *Drosophila melanogaster*. In the space of a few weeks we would open up the cramped rooms and turn the apartment into a glorious living space filled with light.

♦ ♦ ♦

But all that came later. In early 1981 I began organizing a return visit to Venezuela, this time with a team of clinicians and geneticists and a nurse to do the blood draws and skin biopsies we hoped to carry out as part of our mapping project. I knew right away I wanted Anne Young, a spectacular young neurologist and neuroscientist, to be part of the team. By the time she joined the faculty at the University of Michigan Medical School, soon after I left Ann Arbor, Anne had already done pioneering work on neurotransmitters as a medical student in the lab of Solomon Snyder at Johns Hopkins. She would go on to make major contributions to conceptualizing critical functional pathways in that part of the brain called the basal ganglia, among many other achievements. I had met Anne and her husband Jack Penney, also a neurologist and neuroscientist, when I returned to Ann Arbor to give grand rounds and HD families in the region asked me to check her out. I took an instant liking to Anne, with her direct, outspoken manner, her lack of pretention, and her obvious clinical talent and expertise. She was exactly the kind of imaginative, tough, clear-thinking clinician I wanted for my team. She would become my best friend and right-hand person, joining every one of the study visits to Venezuela we made over the following two decades. I couldn't imagine going there without Anne. Jack would later join us, and even their two teenage daughters Jessie and Ellen came on several trips and immediately made themselves valuable workers.

Ira Shoulson, a gifted young neurologist from the University of Rochester, in New York, also joined the team early on, bringing his reassuring manner and wry sense of humor. A wisecracking Argentine former emergency room nurse, Fidela Gomez, had skills at drawing blood

*Anne Young and I in the office at the Hotel del Lago, Maracaibo.*
*(Photo © Steve Uzzell, All Rights Reserved.)*

and making jokes in both Spanish and English that helped ease anxieties and were critical to our success. Although not a clinician, Mike Conneally helped out several times—mostly distributing candy to the local kids, as did other geneticists, neurologists, genetic counselors, psychiatrists, social workers, nurses, postdocs, and graduate students. Even some people who had no special connection with Huntington's, either professionally or personally, came with us on occasion. Although most came from the United States, some came from elsewhere, including

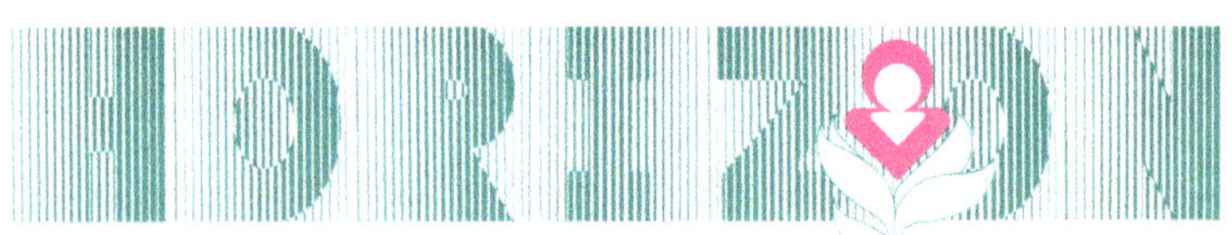

HUNTINGTON SOCIETY OF CANADA SOCIÉTÉ HUNTINGTON DU CANADA

No. 35 Winter 1984

**Dr. Nancy Wexler** (above left) and members of
**Venezuela Research Team**

**The following is the first part of a summary of Dr. Wexler's address to the Annual Workshop Weekend.**
I hadn't intended to begin this way, but since Ralph was so generous, I would like to take a few minutes and be more personal.
Never in my wildest dreams did I think that I would be able to have a part in changing Huntington's disease for the world. My mother was diagnosed with HD in 1967, just after I graduated from college. My family is not a medical one - my father is a psychoanalyst, and my sister is a historian. All of us got together, my father, sister and Maryline Barnard, a close family friend and psychologist, and said, "well, what do we do?" Luckily, I have a very optimistic, creative, courageous and energetic father. He said, "well, let's get it before it gets us, that's what we do". So we started out to do just that.

Since 1970, our Foundation has been having small interdisciplinary workshops, all around the country, bringing some of the brightest, finest, most sparkling minds together to grapple with this problem. Many of these scientists have never seen a patient with any illness - much less HD. Many are PhD's; they work in the laboratory, they work with mice; sometimes with little fruit flies; they work with chemicals; with DNA, the sub-



*Venezuela team with Nancy. (Courtesy of the Huntington Society of Canada.)*

Venezuela, Argentina, China, and Spain. Besides contributing to the research, a whole cohort of future Huntington's disease specialists—many of them women—got valuable training during these expeditions where they saw more cases of the disease in a week than they might see in a month, a year, or even, as some remarked, a lifetime.

For the first 15 years Jacqueline Gray (now Jackson) came with us each spring to help organize the project in the field and provide support. Jackie worked with Mike Conneally at the Indiana University School of Medicine and later helped manage the Huntington's Disease Research Roster that Mike created in response to our Commission recommendation. A few years later a stroke of amazing luck brought Judy Lorimer to work with us in our New York office as an HDF administrator, just when I needed to hire more help: It turned out that she was living near Herb and me with her husband, my high school friend, guitarist Michael Lorimer. After seeing my photo on the cover of a Columbia magazine, they reached out to me, and I hired Judy on the spot. Judy had grown up in Red Lodge, Montana, and had the strength and resilience of her western origins, qualities that served us well in Venezuela as they did in NYC. Another stroke of luck about a decade later introduced us to Julie Porter,

(Left to right) *Judy Lorimer, me, and Julie Porter. (Photo courtesy Hereditary Disease Foundation.)*

who had been captivated by the HD story since learning about it in her high school genetics class. After graduating from Columbia in 1996, she reached out to me at a time when the Venezuela project—and my many other university and foundation commitments—had expanded to the point that Judy needed additional support. Julie immediately showed her gifts for organizing and for working well with everyone, bringing her energy, warmth, and clear-headedness to a job with many complex dimensions. We three became trusted friends and remain so to this day.

♦ ♦ ♦

How to pay for our Venezuela study visits, of course, loomed large from the outset. Sometimes I just got lucky, as when two of my bosses at the NINDS authorized funding for our early trips, even though, as staff in the extramural program, I was supposed to support research outside of the institute, not perform research myself. Without my having to go through an arduous grant application process, Tom Chase had made the 1979 exploratory trip possible; Jack Brinley, another director at the institute who had connections in Venezuela and spoke Spanish, authorized a second visit, in 1980, to secure a contract for collaboration with Venezuelan researchers at the University of Zulia. And Jack came with me to carry it out. (Negrette called it "the most fruitful scientific collaboration ever signed by this country in any epoch."[8]) Two more trips ensued, in 1981 and 1982, without a grant or other official funding.

It was only when Jack Brinley's successor, Carl Leventhal, who was a bit more by the book, became my boss in 1983 and noticed that these expeditions had evidently been paid for out of the "paperclip fund" that things that had passed under the radar started to come to light.

Carl became suspicious. Were the trips to Venezuela a major boondoggle or what! He arranged for a group of eminent geneticists to come to Washington on a site visit, ask questions, and give a report. He probably thought this site visit would result in cutting off the Venezuela trips without his having to do so on his own. Prior to the visit, I was to write a report of our previous trips and what we proposed going forward. Instead of a sober accounting, my report was basically a pitch for why the Venezuela families were so important for finding the Huntington's gene and why building a pedigree and taking samples were key to advancing knowledge of the disease. Leventhal called me on the carpet. My report was too biased, too partial, he exclaimed, and he accused me of "willful insubordination." He told me this could get me fired, which was not a good way to start off a new professional relationship. He also instructed me not to speak at the site visit: I could not present slides or my report; I could answer only yes or no to questions. Of course I burst into tears, fearing that all my hopes were about to be dashed. I doubt he would have spoken that way to a male colleague, but I didn't question him at the time.

High-level geneticists came on the site visit, including Luca Cavalli-Sforza from Stanford, one of the most distinguished human geneticists in the field. Fearing the worst, I was astonished to hear him say he thought it was a wonderful project and could he too get some Venezuelan DNA samples. So, then Carl got all kinds of credit for having a terrific project in his portfolio, and all his threats were out the door.

Most of the time throughout this period, I was dealing professionally with men far older and more experienced than I was. I never forgot that I was a woman in the mostly male world of the NIH, although gender was not front and center in my mind, except on certain occasions, such as at a meeting of the World Federation of Neurology in Oxford, UK. There I was talking up the Venezuela field research to anyone who would listen, and while a few people at the meeting, including HD specialists Ntinos Myrianthopoulos and Ted Bird, were enthusiastic, most of them seemed to think a woman, especially a young woman like me, could never pull off such a large project. It was discouraging and dispiriting at the time. In retrospect, though, I think their sexist attitudes only made me more determined to succeed.

♦ ♦ ♦

Just traveling from New York to Maracaibo every spring was a major undertaking, with the enormous suitcases, duffel bags, and boxes full of the supplies and equipment we brought with us, along with donated medicines, clothing and toys. Our Venezuela research, in our minds, was always a two-way collaboration between our team and the HD families. We were asking them to contribute blood, skin, and, later on, sperm samples, along with their expert knowledge of their family histories, and their own diagnoses of who in the past did or did not have *el mal.*

We were asking them also to undergo neurological and cognitive tests, typically every year, so we could follow the course of the disease. And in return we would offer counseling and basic medical care—for parasites, malaria, infections, fevers, and wounds, along with prenatal vitamins and birth control information and supplies if they asked for it.

Without effective treatment for Huntington's, in Venezuela or anywhere else, we focused on treating symptoms to the extent we could, bringing with us haloperidol—an antipsychotic medication used for Huntington's in the United States, although its negative side effects made it a last resort. But haloperidol was expensive worldwide and often got held up in customs. And it required monitoring and access to continuing care. We tried to interest the medical students at the University of Zulia in coming to nearby San Luis and Barranquitas to see the HD families but without much success. We had more success securing donations of medicines for everyday conditions from various international charities and philanthropic organizations, while the dean of the School of Public Health at Columbia, Allen Rosenfield, helped secure IUDs and diaphragms. But all these were temporary infusions of aid, as we were only too aware.

♦ ♦ ♦

On that first team visit in 1981, we developed a routine that we would follow from then on. As soon as we arrived in Maracaibo, we set up our "office" in one of the rooms at the Hotel del Lago downtown where we all stayed. Then we jumped into rented cars and pickup trucks and made our way to San Luis. The barrio hadn't changed much since my first visit three years earlier, its dusty, unpaved streets still lined mostly with cinderblock houses sporting neatly swept dirt floors and outhouses in back. The community faced directly onto the lake, its beach shaded by tall, skinny palm trees swaying over the *chalanas* parked on the sand next to fishing nets spread out to dry. Once again, Dr. Negrette welcomed us. This time, we learned more about his life and the origin of his interest in *el mal*. After attending medical school at the University of Zulia, he was assigned as a physician to San Luis for two years of public service, as required in Venezuela at the time. Having grown up poor in the nearby town of La Cañada, he was familiar with the Maracaibo region, although not with San Luis. He told us how at first he considered the people there rude and ill-tempered, prone to drunkenness. One day, a little old lady grabbed him on the street and told him he was a bad doctor; she said the people weren't drunk, they were sick. So, he began visiting the houses and noticed that their "drunkenness" ran in families. He looked in medical textbooks to find out what might be wrong with them. He began to feel affection toward them, to make friends with them. When he told his medical colleagues at the university that he thought they had Huntington's chorea, the senior medical professors scoffed. They told him it could not be, that they did not have Huntington's chorea in Venezuela.

*Dr. Américo Negrette, physician who diagnosed Huntington's disease in Venezuela in the 1950s. (Photo by Alice Wexler.)*

Undaunted, Negrette began assembling pedigrees—genealogical charts (family trees) indicating the presence or absence of the disease in each person represented, making it possible to follow the path of the illness over generations. He also made clinical notes and gathered pathology reports to give a wide-ranging portrait of Huntington's in the state of Zulia. A poet as well as a physician, he captured the voices of the people living with *el mal* and their relatives, their sorrows and their strategies for surviving. Negrette presented papers on the disease at medical conferences and, in 1962, published the first monograph on HD, *Corea de Huntington: Estudio de una sola familia investigada a través de varias generaciones.*[9] Eventually Negrette too became a medical professor at the University of Zulia, and some of his students also presented papers at conferences. One student was Ramón Ávila-Girón, who showed the films we saw in Columbus at the 1972 Centennial Symposium that made me determined to come to this place.

When we arrived that spring of 1981, Dr. Negrette again walked with us through San Luis, where we greeted people we had met three years earlier and were introduced to many others. They trusted Negrette and so they trusted us and felt that we were safe for them to know.

Nonetheless, it became quickly apparent that we needed to take more time explaining why we were there and why their participation was so crucial to finding a treatment for *el mal.*

After a few days, we organized a gathering in the home of a local teacher whose house was somewhat larger than most. Soon, there were so many people with Huntington's present that the living room appeared to undulate—shoulders, arms, trunks all pulsing to their private rhythms. Although I'd been practicing my Spanish, I decided that I needed Fidela to translate, to make sure I was getting my points across. I thanked the people for participating in our study and explained how important they were in our efforts to find a treatment and cure for Huntington's because the research needed big families. I explained that my mother had also had *el mal.* I said that I and they probably would be able to confirm that we were related if we could draw our family trees back far enough. I told them I had inherited *el mal* as well (in their vernacular, everyone who has a parent with the disease inherits it but only some develop symptoms). My family was also part of the study. When I confirmed as much by showing them my skin biopsy scar, they were shocked. At first, they didn't believe me. Most of them were under the impression that they were the only ones in the world with this affliction. How could a country that had sent men to the moon tolerate having this disease? Surely, we would have cured it already! They were dubious that my family really had *el mal,* so I was passed around the crowded room in order that everyone could see my forearm, with the small, white, half-moon scar exposed. And then, "She has the mark! She has the mark!"

That was the beginning of our friendship. People were willing to donate blood for the sake of their relatives and for their children and grandchildren. Over time we collected some 4,000 blood samples without anyone really giving us a hard time. I became known as "the blonde with the mark," or "*la catira con la marca.*" Eventually, from the children who ran after our car to the old people drowsing in the sun, everyone called me simply Nancy.

For some critics, however, displaying my "mark" constituted undue pressure on vulnerable human subjects, a form of psychological coercion. I felt that showing off my mark was a way of indicating my own vulnerability, showing that I identified with them, letting them know

*Anne Young, Jacqueline Bickham (a psychiatrist and team member), and me comforting a boy in San Luis. (Photo by Jeffrey A. Szmulewicz.)*

that I was at risk for the same disease from which they were suffering. I, too, had a personal stake in this research. These critics also objected to our paying research subjects for their participation, a common practice in the United States. We felt it would have been unethical *not* to give reasonable compensation to the Venezuelans for their time and effort. They were, after all, taking time off from their own occupations to perform valuable labor for us. Not to mention the fact that they were extremely poor. We felt they had earned the payment they received.

♦ ♦ ♦

Very quickly during that 1981 visit, we established our chain of operations: My team and I would gather samples and pedigree information in the field. We would then send everything to James Gusella, initially an assistant professor in the genetics department at Harvard Medical School. David Housman had recruited Jim, his talented former graduate student, to direct the laboratory work (because Housman was at MIT and the marker project was at Harvard). Jim and his team would then test the samples for a marker and forward their results to Mike Conneally in Indianapolis.

There, Mike and his team of statistical geneticists would run Jim's data through their computers to come up with a statistic about the likelihood of linkage—a statistic known as a LOD score (short for "logarithm of the odds"). A LOD score indicates the likelihood that a result is *not* by chance; in this case, the higher the LOD score, the more likely it would be that the score indicated a true linkage between a marker and the sought after gene.

Though we still hoped to discover the elusive homozygote, we were now putting our energies into studying complete families, the larger the better as David Housman had advised. Using Negrette's pedigrees as our starting point, I began to recruit volunteers, not only those living with Huntington's but also their relatives without HD who were important to the study as well. Some of them were at risk, some had no risk, but every person contributed valuable data. We always tried to look at *all* members of the families, even when it sometimes meant trekking off to distant villages and hamlets and checking out local bars and boats.

Identifying people accurately sometimes challenged our creativity as the Venezuelans we encountered had a very fluid relationship to names. A teenage boy insists he's called Eduardo, while his mother barks from across the room, "Manuel! His name is Manuel!" "No, I'm Eduardo!" he argues. We meet sisters named Maria Carolina and Maria Salina. The next time we meet them, they've switched middle names or first names.

*Talking with kids in San Luis. (Photo © Steve Uzzell, All Rights Reserved.)*

We had a difficult time describing people even to our own team. Yet we knew it was a matter of life and death to get the correct name and family relationship connected to the right DNA sample. If we made a mistake and reported that Maria Carolina has Huntington's and Maria Salina doesn't and we're calling them by the wrong names, the geneticists will never find the gene. And if we couldn't even be certain that we were assigning the two Marias the right place in this ever-growing and complex family tree, how could we assure everyone else that the DNA, clinical information, and family relationships were accurate? Fidela called every woman of a certain age "the little old lady in the hammock by the beach." Because this description fit half the people in the town, we needed a way to determine *which* little old lady we were talking about.

Steve Uzzell, a *National Geographic* photographer and friend of mine whom we hired to accompany us with his camera on the initial trips, had the inspired idea of making black-and-white contact sheets of the Polaroids and pasting them on the pedigree chart. Now when I asked people who were their parents and grandparents and who had *el mal,* only to receive a rapid-fire string of relationships shot back at me in Spanish, I just directed them to find the face of their nearest relative on the wall and stand under the picture. We would point to the picture and ask, "Which is this person to you? ... Oh, your mother's brother's wife's cousin's aunt who had ten children? Okay." And then we would draw new pedigree information right onto the paper on the wall and paste up new Polaroid pictures. Obstacle overcome.

We soon confirmed that most of the HD families in San Luis, Barranquitas, and Laguneta traced their ancestors back to a woman in Laguneta in the early nineteenth century, fittingly named Maria Concepción, although where her forebears originated we never managed to ascertain. Early on we also found a few families whose origin, although undetermined, was completely separate. We came to know all these families well, seeing them year after year for more than two decades. Especially in Laguneta, where we would rent several of the houses on stilts over the lake for three or four days and nights and make one our clinic and the others our hotel while we carried out research activities and offered medical care. I came to enjoy sleeping and even working in a hammock and after a while felt almost at home.

*Pedigree with Polaroids. (Photo © Steve Uzzell, All Rights Reserved.)*

And yet our return to these Venezuelan communities, and to the same families, year after year for more than two decades remained a journey across class and culture. Everything we did had to take into account their poverty, isolation, lack of access to education, and the stigma that outsiders imposed on them. We had seen early on that those outside these communities often treated people with Huntington's as pariahs,

*At work in Laguneta. (Photo by Alice Wexler.)*

*Nancy and Anne Young in Laguneta. (Photo by Alice Wexler.)*

as if HD were contagious, as if the people were somehow contaminated and their communities dangerous places to be avoided. It was as if these outsiders couldn't see the people for the disease.

This attitude always brought back to me a scene I had witnessed in a psychiatric hospital in Shanghai on my trip to China. We had met a ten-year-old girl living in the hospital who hadn't uttered a word in years, breaking her silence only with an occasional outburst of high-pitched, tuneless singing. With an air of dismissive finality, the staff offered their diagnosis of "childhood schizophrenia with mental retardation." I watched as my father approached her and sat down on one of the tiny chairs next to her, right at her level. She scowled and moved slightly away, fixated on a game she was playing. Dad talked to those of us watching and gradually, through our translator, he began to talk just to the girl, joining her, uninvited, in playing her game. She noticed but didn't speak. He hunched in the chair beside her and moved almost nose-to-nose, playfully talking to her. When it was time to leave, he asked her if she wanted him to go. No response. He stood up. She pointed imperiously to the empty chair next to her, silently commanding him to sit down. He did. The nurses begged her to sing. She directed her singing to her newfound companion. He applauded, then stood up again, explaining to her that now he really did have to leave. She said "No," flatly and firmly uttering her first word in years.

Dad looked past her illness, sat at her level, and communicated in a language she could understand, with and without words. And she had responded in kind. I thought often of that encounter when I was in Venezuela and I would try to act similarly, in a Venezuelan register. My father's delicate approach to the nonspeaking girl seemed to me a model of how to communicate across barriers of age, language, culture, and state of mind. My favorite mode of communicating, however, was hugging. People often remarked on my willingness to hold and hug the Venezuelans with Huntington's, not only the children but also the adults, no matter what their physical state. They never seemed repellent to me. Besides, hugging and physical touch come naturally to me. It's my default response, even with strangers, in the United States and everywhere else. The Venezuelans I encountered were physically affectionate. For me, it required no effort to hug them. Nor were my expressions of affection an attempt to coax or cajole them into cooperating with us. They were simply my usual method of operation. They were who I am. Still am.

Another barrier we had to cross was fear of needles for taking blood samples and of small scalpels for tiny skin biopsies. In the beginning, we functioned like a regular clinic, with blood tubes in racks, all sterile, and visible. Of the twenty-five or so people who initially consented, many fainted or threw up. In time, though, we changed how we collected blood. Fidela created a little portable kit with everything she needed, including Batman Band-Aids, wrapped in a Handi-Wipe. Instead of having people come to the clinic, we would visit them in their own houses, where we could explain everything with more leisure and answer all their questions. And everything could be done out of sight; Fidela scarcely needed a needle to draw blood, so skilled was she at this operation. We learned to distract our subjects: one member of our team would hug, sing, or talk to people while Fidela drew their blood. She became adept at drawing blood on moving boats, on top of fishing nets at the docks, in cars rumbling over unpaved roads, in darkened closets, on the hoods of trucks, in hammocks in zillion-degree heat and humidity—and, after the first few, almost no bad reactions.

Somewhere along the line, I decided we should add neuropsychological and cognitive exams to the neurological exams we had used from the start. For my Michigan dissertation research, I had used such tests, which are valuable in picking up cognitive changes that can be the first tiny indication of the HD gene's malevolence. And returning year after year, we were able to test people repeatedly, following those at risk and those with Huntington's prospectively rather than in retrospect, a huge advantage when it came to tracking age of onset and the progression of symptoms. However, the people I was testing had no formal education and were largely illiterate. The tests needed to be modified to make them more environmentally and culturally appropriate. The standard U.S. test of intellectual functioning I had used in my Michigan study, even if translated into Spanish, would not do.

"How are a cat and dog alike?" I asked.

"Enemies," they answered.

"How are red and green alike?"

"Watermelon."

"What month is it today?"

"The month you come."

"How are a horse and a cow alike?"

"Brothers."

The Venezuelans thought in dynamic, interactive relationships, not in fixed, abstract categories. Whether it was a matter of different cultural

*Administering the Purdue Pegboard Test, Laguneta, ca. 1980s. (Photo © Steve Uzzell, All Rights Reserved.)*

lenses, a distinct physical environment, their lack of education, or something else, they related to the world more intuitively than most North Americans taking the same cognitive tests. Their answers made a kind of poetry. Seeing life through their eyes modified my vision as well.

♦ ♦ ♦

When, every year, the time inevitably came to go home, I was always in shock. After living for two months in Venezuela, part of the time working and sleeping in a hammock in a remote stilt village, I would wake up in Washington, D.C., or later New York, suffering extreme culture lag. One day I would be sitting with a family of fishermen in their tiny house over the water, and two days later, I would be meeting with government officials in Washington or with university researchers in New York. My dress and heels felt like a costume. I would look down on my colleagues at home and wonder why they weren't out in the world doing something more useful than arguing around a seminar table. I would still be speaking the language of Maracaibo. Of San Luis. Of Laguneta.

"How are a cow and a horse alike?"

"Brothers."

CHAPTER

# 5

# THE ORACLE OF DNA

I was sweltering in Washington, D.C., in July of 1983, recently returned from our annual team visit to Venezuela, when I decided to phone Mike Conneally in Indianapolis. He told me things were looking good. The LOD score had started to increase for linkage between the Huntington's gene and a marker called G-8 (after Ginger Weeks, a technician in Jim's lab who had selected this probe from a library of probes and prepared and labeled it for use). Everyone in the Iowa HD family *with* Huntington's had one form of the G-8 marker while those *who did not* have the disease had another form.[1]

But Jim and his lab needed to check the samples from more families, especially the Venezuelan families, to see whether that pattern continued to hold. Mike and his family were about to leave for a camping trip in the Grand Canyon, but his lab was on notice, waiting to get more data from Jim's lab to run through the computer. Meanwhile Jim was leaving for a conference in Aspen. (Their ability to go off to a camping trip and conference while awaiting such momentous results is something I—to this day—cannot comprehend. I wasn't capable of straying so far from the phone.)

I called Mike's lab the next day, and one of his graduate students read the computer results over the phone. The LOD score was sky-high. The computer had delivered the news that, as with the Iowa family, everyone *with* Huntington's in the Venezuelan families had one form of G-8, whereas their relatives *without* Huntington's had different forms. The odds of this happening by chance were more than a million to one. I remember letting out a shriek and people running into my office to find out what was wrong. My first thought, after speaking to Mike's lab, was that now I could test myself and *if* I was lucky, I could try to get pregnant. The first call I made was to my father, who answered simply, "I am glad I lived to celebrate this day."

Soon a publicity problem arose. Being in the more open days of the early 1980s, Jim had spoken about G-8 at a recent scientific meeting. Some HD family members attended and, in their excitement, subsequently talked to the press about his findings. An unscrupulous AP reporter, as I consider him, then phoned Joe Martin, threatening to publish the discovery in a newspaper unless the scientists involved made a public announcement right away. He might publish it, the reporter warned, not in the respectable *New York Times* but in a sleazy tabloid such as the *New York Post.* And without confirmation of the facts, there was always a risk of publishing it incorrectly.

We all knew that this discovery was going to be Big News, because it was the first time a DNA marker had been used to locate a disease gene. In other words, the Huntington's gene was the first gene associated with a disease to be mapped to a specific chromosome using DNA markers. And G-8 was relatively close to the HD gene, maybe four million base pairs away. What made it even more dramatic was that the gene could have been on any of the 22 somatic or non-sex chromosomes—that is, almost anywhere among the approximately three billion base pairs of DNA that help make up the human genome.

We also knew that *Nature,* the prestigious science journal where Jim had submitted the paper announcing the discovery, had a strict policy that they would never publish a paper that had been announced publicly beforehand. Joe Martin called *Nature* and asked what they should do. *Nature's* editor essentially told him this discovery is too important to risk the news media getting it wrong. Go ahead and move up the time of the press conference. We'll publish the paper anyway. Joe decided they would have a press conference within a few days in the beautiful Ether Dome, the historic amphitheater at Massachusetts General Hospital where the use of ether for anesthesia was first publicly demonstrated in 1846.

Herb and I were in Dulles Airport outside of Washington, D.C., in the Pan American terminal, when I suddenly heard Nancy Wexler being paged. We were about to leave for India where Herb was invited to give a series of talks. When I got to a phone—these were the days before everyone had cell phones—either Joe or Jim, I can't recall who, was on the other end of the line, telling me they had moved up the date of the press conference, and could I please come up to Boston right away. I had

(Left to right) *Celebrating the marker with Mike Conneally and Jim Gusella, 1983. (Photo courtesy of Heredity Disease Foundation.)*

to choose immediately: Go to India or go to the Ether Dome. I thought to myself, I've never been to India. This was going to be a great adventure and a chance to spend time with Herb. But when we arrived in Delhi, the news about discovering a Huntington's marker was on the front page of the *International Herald Tribune.* I was so furious with myself for not going to the Ether Dome that I threw a wastebasket across our hotel room and bent it all out of shape. To top it off, our luggage got lost for days.

By the time we returned home, news that we had mapped the Huntington's gene to Chromosome 4 had been featured in many mainstream newspapers and on television as well as in the science press. The discovery demonstrated a technique that could be used to locate other disease genes. It was proof of principle for the RFLP approach to mapping genes, which became the basis for the Human Genome Project a decade later. To add to the magic, Jim's group identified the marker on Chromosome 4 on August 4. For all the larger significance of this discovery, though, for me the joy was knowing that, at last, we were closing in on the gene.

♦ ♦ ♦

Earlier in 1983, the Hereditary Disease Foundation had planned a workshop for mid-August in Rochester, New York, to consider the clinical impact of a marker that could predict, with perhaps 95% accuracy, a dreaded disease emerging sometime in the future with nothing you could do to prevent or to treat it. While planning this workshop, we worried that we were being premature, that the workshop would be too "hypothetical"—after all, we expected it to take another five or even ten years to find a marker.[2] Instead, just two weeks before the August workshop was scheduled to start, the news came about G-8 and we found ourselves in a hurry to catch up. We needed to address immediately the fact that *some* individuals at risk for HD could now learn, with a high degree of probability, whether they carried the aberrant gene and would someday develop the disease.

How many people would want to find out? My own PhD research had indicated that as many as two-thirds of those at risk would choose to be tested, motivated by the hope of learning they were free of the unwanted gene or even just to get rid of the uncertainty of 50–50, the haunting probability of 1 in 2. I, too, had assumed that if there were a presymptomatic test—or rather *when* there was a test—I would take it, not least because it would enable me to make an informed decision about childbearing. Now that such a test would soon be a reality, I wasn't so sure. I realized I hadn't thought through all the ramifications, both psychological and social, of what a positive test might mean in my life. And even though I had cautioned in my doctoral thesis that "when the test is actually here, a lot of people are going to change their minds," it never occurred to me that I would be one of them. Now, suddenly, the thought did occur.[3] Even a negative test might pose complexities. After building one's identity around the idea of living at risk, learning that one was free of the aberrant gene could require a dramatic recasting of self. And then, most of the people who tested negative would probably have siblings or other relatives who were not so fortunate. Survivor guilt also lurked in the wings.

Despite Dad's insistence that his clinical judgment told him neither Alice nor I would follow in our mother's footsteps, he started alluding to the curse of Cassandra, the ancient Trojan princess whose ability to see the future was more a curse than a blessing. He was becoming more and more worried about the impact of a positive predictive test if his clinical judgment turned out to be mistaken. And he worried about laboratories undertaking this new technology turning out inaccurate results.

Alice and I planned a weekend meeting at Dad's apartment in Santa Monica with Herb, Alice's then partner John, Maryline, and Dad himself to discuss the test. Things started out badly and grew steadily worse. For starters, Alice and John and Herb and I arrived at Dad's apartment late in the afternoon, very late, so that Dad, who had been expecting us in the morning, was already irritated and upset. And then, very early in our discussion, we realized, much more than before, how deeply entangled we all were in the outcome of the test for both Alice and me. "I feel that Alice's fate and my fate are one and the same," I wrote to Dad afterward. "Both are free or neither are free." Despite our arguments, I told him, "... our family is the bedrock of my life. If I have any psychological strength it is due to this family, and primarily to you ... it is painful when we get pulled apart."[4] And if our family was any example, "people are in for a lot of grief."[5] We ended several hours of arguments about testing without any resolution. Except that I decided to wait until I got pregnant—if I got pregnant—and then test. Wait for a better test, as the science was improving rapidly. Dad acknowledged that it was our decision. Have kids without testing? Dad: "Don't ask me. I know you two, not my grandkids." Dad says if I have kids at risk or that have the gene, we'll work even harder to find a cure.

♦ ♦ ♦

For a while after the marker discovery, the question of whether the mutation causing Huntington's varied around the world remained open—that is, did the gene have different forms, what geneticists call heterogeneity? To answer this question, I traveled to Papua New Guinea, to Peru, to China, to Majorca, and even to Tibet (we didn't find any families with Huntington's in Tibet but we did everywhere else). It turned out that the mutation, whatever it was, appeared to be the same everywhere. That similarity meant that any treatment developed should be beneficial to people with HD worldwide, a huge advantage. It also made the testing situation more straightforward, because the same test could be used everywhere. On the other hand, there was a

5% error rate in the marker test, since it looked for a marker that was a little distance from the gene rather than the gene itself. And that was true everywhere too.

♦ ♦ ♦

The marker discovery also answered a second question: Did a double dose of the mutation produce symptoms different from a single dose? We had started our research hoping that someone homozygous for HD would help solve the fundamental mystery of Huntington's, as it had for hypercholesterolemia. For most dominant disorders, in fact, the clinical symptoms in a homozygote tend to be more severe than in a heterozygote, meaning that even a so-called dominant gene may be only partially dominant. We discovered, to our surprise, that those homozygous for Huntington's were identical in their symptoms and course of illness to those who had just one copy of the gene, making Huntington's the first human disorder to show complete dominance, another unique feature of our fascinating, infuriating disease.

♦ ♦ ♦

I found myself invited to speak at meetings and conferences, especially after I became director, in 1986, of a pilot research project at Columbia on presymptomatic and prenatal testing for Huntington's.[6] I loved exploring the hopes, fears, and dreams of others considering or going through predictive testing, finding in their stories echoes of my own hopes and fears. I was always aware of a certain irony in my situation. I had helped make the marker test possible but my writings about the test were cautionary tales, with titles like "The Tiresias Complex" and "Cassandra's Conundrum" and "The Oracle of DNA." I wanted to challenge suggestions in both the medical and popular media that the way to solve the problem of Huntington's was through presymptomatic gene testing. Such claims, I believed, ignored the psychological burden and potential social costs of living with the knowledge of a future lethal disease for which there was no means of prevention. They also depended on the assumption that those with a positive gene test would not have children, through birth control, abortion, sterilization, or celibacy, options that might not be freely chosen.

We agreed to discuss our thoughts about the test on CBS's *Sixty Minutes* in an interview with veteran newscaster Diane Sawyer. Alice and I were at the age when symptoms often emerged and, so far, no one had suggested that anything was amiss with either of us. Dad saw no reason for hastening the arrival of bad news. Diane Sawyer, however, saw it differently.

Dad: "Well, if you look at these two young ladies and see how they are now ... I know they're at risk, but as far as I'm concerned clinically, I think they're quite safe."

Diane Sawyer: "Are you just wearing blinders? Unwilling to face the reality that you might have to face? Is it a kind of self-deception?"

Dad: "I don't think so. I think it's just measuring risk and reward and saying, 'If they have a 50–50 risk and they both take the test, why would I want to have that risk–reward kind of ratio where one would come up with a bad answer and all three of us are destroyed by it?'"

I knew Dad was speaking out of fear. He was telling us *he* would be destroyed. I had to keep reminding myself that this was *our* decision to make, not his. And yet, we're all in this together...

♦ ♦ ♦

Predictive and diagnostic linkage testing for the marker began being offered in Europe, North America, Australia, and New Zealand in 1986, mostly at academic medical centers. These early tests took place as part of a research protocol to study responses to this fraught information in order to develop more specific guidelines. However, setting up testing in Venezuela, the country that had made the test possible, proved to be much more complex. On our next annual visit after the marker discovery, we organized a meeting in San Luis to explain it to the families. I held up a map of the world and told them it was as if we had been searching for a killer who could have been anywhere in the world, and we'd found the town where he's hiding out, so now we can go door to door. They understood. We asked if they would want to know ahead of time if they were going to get *el mal.* One man said, "If you told me I was going to get the disease and there was nothing I could do about it, I would jump right off that pier out there and kill myself." His words have haunted me ever since. Another man said, "I think it's terrific for myself and for my kids. But I need some help in dealing with this, somebody to talk to." Still others seemed dubious that knowing was even possible because to them it seemed like magic: How could you possibly know the future? Are you a witch?

We went to the University of Zulia in downtown Maracaibo to talk to faculty members about offering clinical gene testing for those at risk. We were in Venezuela only two months a year at most and we did not have the resources to provide such testing, which is usually a process over several months and requiries long-term follow-up. But we were committed to the idea that those who made the research possible should be among the first to benefit from that research. We were chagrined to find that some of the local clinicians and counselors appeared to regard this test as a means to identify candidates for sterilization. We didn't feel confident that testing would be voluntary and confidential, or that the requisite counseling would be given before offering people the test. In addition, abortion was illegal in Venezuela and birth control hard to access, although we offered supplies when asked. We also tried, through the School of Public Health at Columbia, to get Planned Parenthood to offer services locally, without success. What options would people have? It wasn't clear to us how local providers could ensure that presymptomatic testing would be a benefit rather than a harm.

A related question was whether we should provide research participants in Venezuela with the results of individual genetic tests performed as part of the research itself. This practice is not generally followed in the United States because of the uncertain meaning and accuracy of individual results in ongoing research, although it is becoming more common. We felt strongly that disclosing gene status to asymptomatic individuals, especially in an environment of poverty and outside of a supportive clinical context, would not be a "benefit" to these individuals, particularly if local clinicians and others sometimes framed testing and gene status in a eugenic context of disease prevention rather than as a medical choice made voluntarily and free from coercion. In the early days of our work, some Venezuelan scientists and clinicians, including Negrette, openly advocated sterilization as the only solution for Huntington's in that region, a position Negrette subsequently disavowed.[7] We worried that those known to carry the HD mutation could be subject to increased discrimination, targeted as

"San Viteros" even before the first symptoms appeared. We decided that we would give research participants clinical diagnoses when they asked. We would offer genetic counseling and birth control information and supplies when these were requested. But we wouldn't otherwise tell people at risk for HD but with no symptoms whether they had the marker, or later the gene, associated with the disease.[8]

♦ ♦ ♦

Shortly before the deteriorating political situation in Venezuela shut down our annual visits, Columbia University's Institutional Review Board (IRB) threw us yet an additional challenge. They asked that we develop a questionnaire to assess each research subject's knowledge of "safe sexual practices and reproductive options," including prophylactic methods to prevent pregnancy. Such a questionnaire was not required for genetics research in the United States, as far as I was aware. And we had made great efforts to accommodate women when *they themselves* asked for birth control information and supplies. But we felt that to tell a woman she has HD and at the same time that she might choose not to have children was "doing extreme harm... To start suddenly quizzing family members, in a structured questionnaire, if they know how to prevent babies is insulting, intrusive, and likely to be interpreted by them as eugenic, implying that we do not want them to have babies." Although many women did in fact want birth control in some form—usually women who already had children, often women who had the disease themselves—in this Catholic country we felt that *advising* them not to have children violated both their religion and their humanity. I found the IRB's request insulting not only to the Venezuelans but also to me. They were telling *me* that I should not have been born, that my mother should not have given birth to *me*. I replied that if the IRB required the inclusion of "birth control information or birth control supplies as a prerequisite for research on a human genetic disease, this is setting a dangerous eugenic precedent."[9]

Meanwhile U.S. scientists, including Housman and Gusella, were being very cautious about the idea that a marker test should be available commercially. Everyone, including HD family members, wanted guidance. Soon after the marker discovery, the various Huntington's organizations got together to establish guidelines.[10] In my memory, this group was effective because people respected each other, and the professionals realized they had to take their cues from the family members.

And up to now, and after all the intensive, agonizing discussions, only about 10%–20% of people at risk for Huntington's worldwide have opted to get a predictive gene test. Alice and I are not among them.[11]

CHAPTER

# 6

# WILL THE CIRCLE BE UNBROKEN?

I had wanted to have children ever since I was a child myself, lining up my dolls and stuffed animals on my pillow, in Topeka and in Pacific Palisades, so that there was scarcely room for my head. I would crowd them, one after another, and the ones who didn't fit on the pillow I would line up next to me so that we were all nestled together. I loved taking care of little creatures, whether they were animate or inanimate, dressing them, bathing them, feeding them, hugging them, and generally assuming the role of a mommy. While my sister had no use for dolls and never played house, I was the opposite. Being a Mom seemed to me a fine occupation even though, at a fairly early age, I knew I wanted to become a psychologist like my father as well.

When Mom was diagnosed with Huntington's and Alice and I learned about our 50% risk, we both immediately felt we should not pass the disease to any future generations. We were alarmed, not so much by the disease, which we had seen only in our Uncle Seymour's jerky movements and unsteadiness, as by our father's gruesome descriptions of it. Dad couldn't help conveying his own grief and guilt about what he had read to be a truly terrible affliction from Day 1. I don't think he said anything explicit about not having children. But Alice and I both felt we had been given that message, consciously or subliminally. Certainly, that was the advice most neurologists who knew anything about Huntington's gave to their patients up through the 1970s. Even neurologists critical of eugenic sterilization in the United States made an exception for those with a possibility of developing Huntington's disease.[1]

With his huge sense of responsibility, Dad would joke from time to time that if we had kids, he would just have more people to worry about.

Not much encouragement to have families of our own. He loved children and knew how to engage with them, not just with Alice and me, but with our friends, and the kids of his friends and of his patients, and, well, pretty much kids in general. I feel certain he would have adored having grandchildren. It must have been painful for him to listen to tales told by his friends of their marvelous grandkids. As a therapist, though, he also acknowledged from time to time that some grandkids had serious problems and caused more pain than pleasure to those who loved them. Whatever his feelings, he never complained to us about not having grandchildren of his own. As for Mom, she didn't weigh in one way or the other. I'm not sure how she felt. Until the summer of 1968, I looked forward to having kids of my own, although without any sense of urgency.

Mom's diagnosis changed everything. Suddenly we were facing a completely different situation. I was twenty-three, just starting graduate school. Plenty of time to hope for a cure, or at least a treatment. Too busy, anyway, to think about kids just then. Too many boyfriends, too much to do. I kept putting off thinking about it.

My experience in Venezuela intensified my conviction that I should not have biological children of my own unless I could be certain I would not pass on the disease. As hard as it was to see young adults, in their thirties and forties, developing the symptoms, it was that much more difficult seeing so many of the beautiful, bright, lively Venezuelan kids start to slow, become stiff, and lose their vivacity. I didn't want to subject any child of mine to such a fate, although it was true that most of those with juvenile HD had inherited it from their fathers. Nonetheless, the possibility terrified me.

And then I met Herb. I was thirty-six when we met, so the time for childbearing was ticking away, and adoption was difficult, if not impossible, for those with Huntington's in the family. Unexpectedly I got pregnant (just before the G-8 marker was identified). With deep reluctance and many tears, I decided to have an abortion, donating the tissue to what is today the Coriell Tissue Institute for Medical Research in New Jersey. But the first in vitro baby had been born in 1978, and surrogacy would emerge in the 1980s. And after 1986 we had the marker test for Huntington's, although the linkage test required many relatives. The options were widening. I began thinking about trying to get pregnant

*With kids in San Luis. (Photo © Steve Uzzell, All Rights Reserved.)*

again and using prenatal testing to assure a disease-free fetus. I also spent a lot of time weighing the risks of testing myself.

Then suddenly I was pregnant again and waiting for a chorionic villus sampling test for the HD marker, as well as for other problems my age might create.[2] Within a short time, the chief of sonography at Columbia University Medical Center informed me that the fetus had died with an extra Chromosome 13. The third time I found myself pregnant, a Chromosome trisomy 18 again doomed the fetus. We then tried using a donor egg, with Herb's sperm, but those efforts, too, ended in miscarriages. Eventually a new technique then called preimplantation genetic diagnosis, or PGD (now more often referred to as preimplantation genetic testing or PGT), made it possible for a person at risk for HD—or any serious genetic condition—to have in vitro fertilization without having to learn whether any of their embryos inherited the condition and without having to test themselves: Only those embryos that did *not* carry the mutation would be implanted, and the parent at risk would not be informed of the status of the rest. I went to a brilliant and experienced reproductive endocrinologist at Columbia, Zev Rosenwaks, who was renowned as a pioneer with in vitro pregnancies. Unfortunately, his expertise did not work for me.

*With the extended Pardes family.*

At that point, I began to consider whether adoption might be feasible for me, although Herb wasn't interested in adoption. Unexpectedly, I got pregnant again, in Paris, seemingly an augur of good luck, but this time, too, the fetus had a trisomy and died. Altogether I experienced six miscarriages, probably all due to genetic abnormalities relating to my age or Herb's, none having to do with Huntington's disease.

I feel grateful for the opportunities and adventures that have come my way and the incredible people I have had the good fortune to know. But not having biological children of my own remains a sorrow, a grief I continue to live with to this day. Yet it's also true that I consider all the Venezuelan kids we got to know, especially those with Huntington's, as my beloved children. And all of Herb's kids and grandkids. And those of Anne Young and Julie Porter (and Julie herself)! And all the young investigators whom the Hereditary Disease Foundation supports. I love them all. So, in a way, I have many kids to celebrate even if they aren't, strictly speaking, mine.

CHAPTER

# 7

# GENE HUNTERS

"I want to form a group whose goal is to map out this program as it develops." David Housman was speaking on a Sunday afternoon in Santa Monica in January of 1984, in a large, sunny conference bungalow belonging to the Fairmont Miramar Hotel, a place with a casual Southern California vibe, across the street from a park overlooking the Pacific. The Hereditary Disease Foundation's Science Advisory Board was holding its first meeting after discovery of the marker, to consider how to move forward.

It was obvious that many different skills and techniques would be needed to capture the gene. And the technology for isolating it didn't yet exist. The only tool we had was the recombination event, which classical geneticists had long used to figure out relative distances between genes. Recall that recombination is a kind of do-si-do between two chromosomes—the "penguins on an ice flow" exchange of DNA between two chromosomes. And that calculating the frequency of recombination events between two genes or between a marker and a gene can give an estimate of the distance between them on a chromosome. It's a way to physically map genes.

I used to tell people what we were doing was like flying to the moon while inventing the rocket. But we had this hubris that because it had taken just three years to find the marker, we would also find the gene in three or four years. We only needed to find the right people to help, especially those with the latest technologies at their fingertips. And the investigators would have to come from different disciplines: biochemistry, biophysics, molecular biology, molecular genetics, and neurobiology, among others.

For several years, the foundation had been pursuing more outreach than was the usual practice among grant-funding organizations, which typically put out a call for proposals on a topic and wait for applications to come in over the transom. Our then–scientific director Allan Tobin and I were proactive, attending as many science meetings and conferences as we could to meet people and keep up with the latest developments. Now we were making outreach an explicit policy: We would seek out people who had new technologies—many of them young investigators early in their careers—and invite them to try their approaches on Huntington's. We had always seen the foundation's role as that of a catalyst. As a small foundation with a limited budget, we couldn't match funding from the government or big foundations, but we could offer seed money at an early stage for projects to get rolling, and sometimes for pilot projects that were risky. We wanted to be a catalyst in ways that the NIH couldn't be.

There were also many questions about use of the G-8 probe—who would have access to it, and how it would be distributed? Jim Gusella had led the laboratory effort searching for the marker and he had the G-8 probe in his lab. He wasn't eager to hand it over to others. But by his own admission, four million base pairs away from the gene was a considerable distance, even if both were near the top of the short arm of Chromosome 4. There was also the question of whether the foundation should support clinical research—for instance, on the impact of the predictive test.

*With David Housman at the whiteboard. (Photo by Jeff Szmulewicz. Courtesy of HDF.)*

It was an historic moment, a dangerous moment according to some. We'd taken an audacious approach, and it had led to a spectacular result. And now we had this new tool, this G-8 probe, but also a probe with immense social and emotional power, capable of liberating some of those at risk from anxiety and despair and possibly condemning others to a lifetime of dread. What were our responsibilities as investigators, as a foundation? What were my responsibilities as a researcher, as foundation president, as a person at risk?

Dad felt strongly that the foundation should fund basic and not clinical research. His donors, he said, were interested in supporting science that could impact diseases in *their* families. They would be less inclined to support research aimed primarily at Huntington's. But science board members felt that we had to take responsibility for the outcomes of the marker discovery. That included studying the clinical impact of presymptomatic gene testing, as this type of definitive test predicting the future had never been done before in medicine.

♦ ♦ ♦

A few years earlier, we had started to use collaborative agreements as a cross between a traditional investigator-initiated research grant and a contract for achieving a specified aim. The idea of collaborative agreements had been suggested to us by two science board members, Ed Kravitz and clinical psychologist Steve Matthysse, who like Ed was at Harvard Medical School. Both Ed and Steve had been instrumental in shaping and sustaining a cooperative ethos and in recruiting innovative thinkers. Now, with the challenge of moving from the marker to the gene, collaborative agreements were more necessary than ever. And so, right away at that first January meeting after the marker discovery, we began to set up ground rules. At first, we created several small agreements among just two or three labs to perform a specific task. Over the course of several years, we developed one large collaborative agreement among seven and then six labs. These labs agreed to share resources, reagents, assays, information, and, most critical of all, credit.

Over time there was some changing of the guard but most of the principal investigators (PIs) stayed, each bringing their special skills and experience along with their teams of postdocs, graduate student, and technicians. From MIT came David Housman, architect of the mapping project; from Harvard came Jim Gusella, mastermind of the G-8 marker with Marcy MacDonald; their lab would become the hub and headquarters of the project; John Wasmuth joined from the University of California at Irvine, along with his postdoc Leslie Thompson; Hans Lehrach joined from the Imperial Cancer Research Fund in London, with his postdoc Gillian Bates; and Peter Harper came from Cardiff University in Wales. Several years later, Francis Collins joined from the University of Michigan.[1] We offered each of the PIs $30,000 annually (equal to $90,000 today) to fund lab costs and personnel.

The bedrock of the group's cooperation, which we arrived at within the first couple of years, was our agreement for collective authorship on the paper announcing the gene discovery: the Huntington's Disease Collaborative Research Group. Working together to resolve what could have been a deal breaker proved to be a springboard for bonding. Although physicists have long worked in large collaborative groups, this level of cooperation was so unusual in biomedical science at the time that one member of our scientific advisory board deemed it "a scientific experiment in itself."

Four times a year we met to share our findings, most often at the Miramar, where we continued the freewheeling style of the workshops. We found that plucking everyone off their hamster wheels and transporting them to a pleasurable setting freed their minds. And unfettered imagination was what had taken us so far so fast. Although sometimes people complained about having too many meetings, everyone agreed that coming together in person was essential for good communication in those days before email and Zoom. (It was only in 1986 that the foundation began paying $10 a month for participants to get "electronic mail" and a "mailbox.")

At those early meetings we discussed how to divide up tasks, who should do which experiments, and how to make sure those who committed to doing them actually carried them out and shared the results with everyone. We talked about how to make sure postdocs and junior investigators working on projects and experiments got credit for them and felt part of the group. We mulled over whether those who could do experiments fastest should have priority or whether there were other considerations, and if having experiments done in two places might be okay or even desirable at times. And we debated what to do when, say, Cambridge and Cardiff got different results from the same experiments. We wanted to avoid massive amounts of overlap, but we also wanted to keep people enthusiastic and involved, including the postdocs and graduate students who did much of the hands-on work.

Although conflict was minimal, one issue that came up repeatedly at our meetings, especially in the early days, was the extent to which rules of the collaborative group should be spelled out explicitly, which I supported. So did Dad, who came to the first meetings of the group and helped set the cooperative tone. But many participants opposed this route, being accustomed to working more informally only with their own lab members or in much smaller collaborations. In any case, the emergence of rapidly evolving technologies made the need for teamwork obvious. Individual investigators understood from the start that they couldn't really work independently; they needed their colleagues' skills and techniques. Working often with data from sources other than our own, the Gene Hunters, as we began calling ourselves, learned to speak the same language, coalescing into a tight community.

♦ ♦ ♦

One of the great benefits of collaboration is the balancing of moods. When one is down, another is up. At meetings I would try to read the room, finding pockets of positivity when people were feeling frustrated and using them to raise everyone up. At the same time, I didn't discount anyone's negative feelings. These were skills I developed growing up in a family where my life depended on it. And I felt appreciated and validated by my scientist colleagues, whose generosity and determination kept me going every day.

I had learned another valuable skill from my family: how to act in the world as if everything were okay even when it wasn't. I used to give spontaneous speeches at our meetings, like a cheerleader. September 1986, Sonesta Hotel, Cambridge, Massachusetts: "When we first started out in 1968," I told the group, "no one had heard of Huntington's disease. The distance we've come is captured by someone saying recently that it was difficult to work on Alzheimer's because it wasn't as sexy as

Huntington's. That put it in perspective. That's largely because of results coming out of the workshops. My father had the idea that standard ways of doing science—though often effective—would not be effective for a relatively obscure disease. We've tried to make a tradition of not doing things in the traditional way. We're trying to get the gene but also to do it in an imaginative way, a harmonious way, because we thought that was the best way. 'Gee, so and so has the cell line, do you think I could get it?' Now that method has worked. The question is, can we continue in the same way? There's a lot more at stake now because you've been so successful. It would be helpful to confirm this collaboration. If we're going to change the collaboration, we should say so now." And so on and so on.

Ironically, the year we discovered the marker, the federal government cut back funding for the NINDS, although fortunately not ours. The Reagan revolution was in full swing, with tax cuts for the wealthy and cuts in funding for science and social services. I was determined that our team would not be distracted from their efforts by financial concerns. Fortunately, the combination of funding from the NINDS, the Harvard Center Without Walls grant, and the Hereditary Disease Foundation sustained the collaboration, along with another significant contribution from an unusual source.

Dennis Shea was a bonds broker from New Jersey who had changed his career focus from high finance to disease advocacy when he discovered Huntington's in his family. In addition to enlarging our coffers, Dennis boosted our morale by sponsoring informal meetings for the Gene Hunters every spring at his resort compound on Islamorada in the Florida Keys. His generosity enabled everyone to gather for one-on-one interactions in a sublime tropical setting.

Besides being able to discuss experiments, we could also diffuse any tensions that had arisen. Within a few years, the collaborative had grown to include fifty-eight researchers and, at times, as many as ten institutions. The more moving parts, the greater likelihood of a breakdown, and I felt it was my responsibility to make sure no one was left stranded at the side of the road. Possibility could not be allowed to die in disagreement. A case in point: the conflict of two groups butting heads over wanting to do the same experiment. My intervention could be perceived as favoritism. So, what if both groups agreed to do the same

*Huntington's Disease Collaborative Research Group and members of the advisory committee at Islamorada, Florida.* (From left to right) *Standing: me, John Wasmuth, Marcy MacDonald, P. Michael Conneally, David Housman (behind Conneally), Francis Collins (behind Housman), Sian Searle, James Gusella (in glasses), Julia Richards, Allan Tobin, Michael Altherr* (in back)*, Hans Lehrach, Robert Moyzis, Gillian Bates, H. Robert Horvitz, Glen Evans, and Herbert Pardes. Seated in front: Lynn Doucette-Stamm, Anna-Maria Frischauf, Jan-Fang Cheng, and Norman Doggett. (Courtesy of the Huntington's Disease Foundation.)*

experiment? We might, in fact, learn more from looking at the problem two different ways. Conflict often begat progress.

The relaxed Islamorada meetings also helped build collegiality and friendships. We swam, went for boat rides, lounged on the beach, even climbed a palm tree, as Hans Lehrach famously did. And always talked science. We also had the benefit of our awesome advisory committee in attendance—Steve Matthysse, Glen Evans from the Salk Institute of Biological Studies in La Jolla, California, geneticist Richard Mulligan, a future MacArthur "genius" award recipient from Harvard, and MIT's Bob Horvitz—a high-power group offering guidance and perspective. Besides his skills as a scientist, Bob brought with him the ethos of openness and sharing for which worm biologists were especially noted. His calm voice and the thoughtful way he offered creative solutions helped move us through some choppy waters.

♦ ♦ ♦

Things were going remarkably well, with participants even bringing their probes in tiny tubes—occasionally packed in ice cream cartons!—to hand out at meetings and generally being open and forthcoming. I remember Steve Matthysse saying, three years in, that the atmosphere we had created so far was just what it ought to be. And the highly critical Richard Mulligan exclaiming that what we were doing was "fantastic" and that he couldn't think of doing anything differently. And John Wasmuth, also not given to hyperbole, exclaiming that it was a bit staggering how fast things were going. I, too, recall feeling confident that, in 1986, we were "coming into the home stretch."

♦ ♦ ♦

The Gene Hunters' first achievement was to narrow the location of the Huntington's gene to the top 3% of Chromosome 4, a region of some twenty million base pairs, extending from the G-8 marker to the chromosome tip, called the telomere. And then came our first big break. A "flanking marker" was found; in other words, a marker on the *other* side of the gene from G-8. The flanking marker was closer to the chromosome's tip and further defined the target region. We were closing in! As early as January 1987, just three years into the collaboration, the consensus was that we would find a cloned piece of DNA containing the HD gene in less than a year.

But almost as quickly as the door had opened, it slammed shut. The only real discovery here was the supreme importance of accurate clinical diagnoses. It turned out that the person on whose DNA the "discovery" of the flanking marker was based probably didn't have the disease after all.

The Venezuelan data suddenly turned up DNA from a pair of siblings that suggested two possible locations for the Huntington's gene on Chromosome 4. One sibling's DNA placed it far out on the tip, consistent with our prior findings, whereas the other's put it more toward the center of the chromosome. Someone raised the possibility that the Huntington's gene might be a piece of another chromosome displaced onto the top of Chromosome 4. Or a part of Chromosome 4 that was usually inherited along with the Huntington's gene might have switched places with part of a normal chromosome, leaving only a tiny segment out at the tip containing the gene itself. When postdoc Gill Bates, in 1989, cloned the tip of Chromosome 4, we felt on the brink of solving the mystery. But then she found that the tip of Chromosome 4 from a person *with* Huntington's appeared identical to that of someone who *did not* have the disease.

Despite these setbacks, we continued to feel optimistic. Each January, we would predict that the HD gene would make its appearance before the end of the year. And then December would arrive, without the gene. Part of what sustained us was the feeling that the collaborative agreement itself was a significant accomplishment, with its free flow of materials and information among all the participating labs. Everything was working well. Except that the gene still eluded us.

♦ ♦ ♦

Six years into the search, members of the group were growing tired. I recall Glen Evans telling us, in January of 1989, that when he walked into that morning's meeting an hour late, he could feel something wrong, he could feel the tension in the room. He thought any newcomer would think twice about joining the effort. Not good if part of the foundation's work was to attract new people to Huntington's research.

And yet others continued to feel the achievements vastly outweighed the problems. I wrote down all the praise so I could remind myself when, on occasion, my spirits started to sag: John Wasmuth noting that this was the first group ever to travel in detail down a chromosome; Francis Collins crediting the group with establishing a paradigm for mapping the human genome, a project that was in the planning stages and which he would later lead. We were doing fine, according to Mike Conneally, working more effectively together than collaborations he was involved with for different diseases.

Of course, there was always the nagging worry that other researchers would get there first. Although those inside the collaboration remained friendly and cooperative with those outside, competition is competition. Underlying any frustration, disappointment, and sense of urgency was the fact that for some of the researchers, joining the collaboration came with considerable risk to their career. A participant might be responsible for a groundbreaking discovery but the credit, as agreed to beforehand, would be shared by the group. I especially tried to look out for the young postdocs, to make sure they got the credit they deserved and to encourage and nurture them in any way that I could. I would get to know them and build relationships with them, find out about their dreams for the future, and sometimes even about their family crises back home. It wasn't a calculated strategy to keep them working on Huntington's; it came naturally. I loved them and I loved mentoring them. I cared about their lives. They were my adopted family too.

♦ ♦ ♦

There were still many questions: How close was the gene to the telomere—the tip of the chromosome, where some evidence still placed it. This telomere was proving to be an inhospitable terrain full of repeated DNA sequences and other hard-to-clone material. It resembled a jigsaw puzzle, according to John Wasmuth, where a lot of the pieces looked alike and it was hard to see how they fit together. Other evidence pointed farther down, toward the more internal region of the chromosome, an area vastly larger to search. But at the same time, *if* the errant gene were located in that internal region, the problems it caused might be more fixable. A setback and breakthrough combined.

We worried about different complicating possibilities: What if the gene was only transiently expressed, early in development, and then turned off for the rest of a person's life? Were neurons born defective or did they eventually die due to what was happening around them? Did cells in the brain die because they were programmed to die, or were they killed by something outside? Were we seeing a case of murder or suicide?

Then a new concern arose that the havoc wrought by the gene might be set into motion in early development, planting a time bomb that would go off twenty or thirty years later. Perhaps a normal rearrangement that was supposed to take place during development didn't occur. If that were true, we would have to be time travelers to solve the problem. Then there was discussion of an aberration in which a normal gene is altered because it gets switched to some abnormal position on the chromosome. Or perhaps a suppressor gene, whose job is to turn off another gene, fails somewhere along the way—anywhere from birth to onset of the disease—and whatever was supposed to be suppressed goes on unchecked. And most vexing was the possibility that there could be double recombinants or shadow recombinants pointing in the wrong directions. We wondered if perhaps we should change our name from Gene Hunters to Ghost Busters.

All these possibilities continued to haunt us. To the point that several people began expressing skepticism about how much the gene would actually tell us about the disease. Even with the gene in hand, we would need more physiological studies. Anne Young, without whom I couldn't imagine doing any of this work, kept reminding us that the foundation needed to fund other research besides mapping the gene. She said waiting for the gene was a bad idea and we agreed and continued to hold workshops on topics other than the gene.

One thing we all agreed on: As the years ticked by, it never got easier. We had cracked open a universe where discovery bred discovery and yet the gene remained beyond our reach.

♦ ♦ ♦

Certain discoveries are conceptually critical, like the double-helix structure of DNA. There is before and after that discovery, which completely changed the way we think. The 1991 identification of the gene for Fragile X, a condition of cognitive disability, was one of those discoveries. The distinctive thing about this mutation was that the gene got much bigger from one generation to the next. This expansion thereby seemed to explain something called anticipation, which refers to the phenomenon of a genetic condition appearing at younger and younger ages in consecutive generations. This feature of Huntington's had attracted the notice of geneticists for a long time, because ~10% of patients were under the age of twenty, although their parents and grandparents had developed symptoms in their thirties or forties. The expansion of the Fragile X gene suggested that the Huntington's gene might expand in a similar way.[2]

Many geneticists had thought anticipation was merely an artefact, but Mike Conneally believed it was a real biological phenomenon.

Even before the Fragile X discovery, I favored Mike's perspective because he was arguing from what he observed in front of his own eyes in Venezuela and that made sense to me. I, too, observed many kids developing Huntington's in Venezuela, more in fact than we observed elsewhere. We could see in the pedigree that over the generations, the disease seemed to appear at ever younger ages. It certainly looked like anticipation. The families even observed it themselves.

But was gene expansion the explanation? These questions soon led to an intriguing new line of inquiry. Years earlier Mike Conneally had noticed that most of those who developed the disease as children had inherited it from their fathers. And even though they had the same gene, these children looked different from their parents. While the parents had chorea, the kids became stiff and slow. We still didn't have an explanation, but the Fragile X discovery of a gene expanding from one generation to the next pointed us to a new line of questioning that would soon lead to some memorable moments for researchers and family members alike.

♦ ♦ ♦

That same year, Kenneth "Kurt" Fischbeck, a scientist then at the University of Pennsylvania (later at the NIH) who would later join the foundation science advisory board, identified the gene for spinobulbar muscular atrophy or SBMA, also called Kennedy's disease, a condition of muscle weakness starting in childhood. This gene turned out to be a small, expanded stretch of three nucleotides, specifically of CAGs that spell out the amino acid glutamine, that repeated themselves a few more times than was typical for boys without Kennedy's. In other words, in boys with Kennedy's disease—and they were mostly boys—there were too many of these CAGs in the androgen receptor gene at one end of the X chromosome. Above a certain threshold, those extra CAGs created big problems. So that discovery, too, suggested possibilities to us.

Then a year later, David Housman and his team isolated the gene for myotonic dystrophy, the most common form of muscular dystrophy. And even more than Fragile X, it turned out to contain an enormous expansion, sometimes as many as several thousand trinucleotide repeats.

So here were three diseases with an expanded trinucleotide repeat. At least two of them seemed to show anticipation. So, again we thought, why not Huntington's? We began thinking that maybe, just maybe, this could be our story too.

CHAPTER

8

# "THE CROWN JEWEL"

As always, I was living in two worlds, but the contrast between them was growing more stark. I was a professor in New York City at one of the most prestigious medical schools in the country, with a spacious corner office overlooking the Hudson River. Soon I would become chair of a working group of the Human Genome Project. I also presided over the Hereditary Disease Foundation, based in Los Angeles. After I left government employment (and could engage in lobbying), Dad had been happy to let me become foundation president, while he became chair of the board of directors, although at first little changed. But foundation activity gradually shifted to New York and eventually the entire office moved. In 1989 Herb added dean of Columbia's medical school to his job description, and ten years later, he would become president and CEO of NewYork–Presbyterian Hospital. We got busier and busier, attending the receptions, benefits, and dinners with donors almost nightly that were part of Herb's job. In addition, Herb always carved out time with his brother, sister, sons, niece, and nephew and soon their partners and eventually his grandchildren. We took many memorable family vacations, making expeditions to places such as Australia, Colorado, and the Caribbean. My life depended on Julie Porter's artfully designed schedules with major events in bold type highlighted with Magic Marker of different colors.

But for six or eight weeks each spring, I immersed myself in three of the poorest communities in Venezuela. And I had many more weeks of Venezuela immersion, if you add the time for preparation before we left for Maracaibo and the time it took when we returned home to decompress and ease back into our New York lives, always a difficult transition for me. I remained haunted by the poverty and isolation of the Venezuelans with whom we worked. We were painfully aware that, in between our visits, these communities lacked basic medical care, even though, in theory, San Luis and Barranquitas had a government-assigned physician and a clinic. Often the doctor didn't show up. Or he'd spend a few hours in the morning before leaving for his private practice in the afternoon. The local physicians never made house calls, despite many of the patients being bedridden or confined to hammocks or a mattress on the floor. There was one exception: a doctor priest, Father Mosqueda, who came to the clinics faithfully and loved the patients. Unfortunately, Father Mosqueda didn't believe in birth control, nor did he understand the genetics of *el mal*. He thought the answer to the illness was for the families to "marry out," a practice that only spread a dominantly inherited disease such as Huntington's in which just one affected parent could transmit it to the next generation.

We decided to look for a doctor who could work with us while we were in Venezuela and who could remain year-round treating the patients. Through a friend of our Argentine nurse Fidela, we met Margot de Young, a Venezuelan physician who had recently completed her medical training after raising two daughters. When Fidela's friend asked Margot if she could recommend someone to work with us and to serve the families in San Luis, Margot replied that she couldn't suggest someone if she didn't know what she was recommending them for. So, she got in her car with her chauffeur (in Venezuela at that time even middle-class people could often afford a chauffeur) and went to the barrio, where she walked through the streets and was devastated by what she saw. She went back every day for a week and came away crying each day. Finally, she said to Fidela and me, "The people there are so warm and loving and affectionate, with no one to take care of them. I would love to do it." Her attitude, like that

*With Margot de Young, ca. 1993. (Courtesy of Maria Ramos.)*

of Negrette, differed totally from the more typical attitude of the local physicians at that time, who regarded San Luis and Barranquitas as dangerous places, with people to be avoided.

And so, in 1988, Margot began working with us. She knocked herself out for the families and they adored her. I adored her. Having her join our team was the best thing that ever happened to us.

♦ ♦ ♦

While the Gene Hunters continued working in their labs, the U.S.–Venezuela Huntington's Disease Collaborative Research Group, as we now called ourselves, continued our work in the field. Starting in 1986, we received three successive five-year grants of one million dollars each from the NINDS and, later on, grants from the W.M. Keck Foundation and from generous individual donors, all of which I administered through the Hereditary Disease Foundation. Each spring I would again lead a rotating team—some repeat visitors and others newbies—to follow the progression of the disease and offer whatever medical care that we could, along with contraception when we were asked.

Soon after Margot joined us, she came up with a brilliant plan for providing more continuity of care. The wife of the governor of Zulia was Margot's friend and together they persuaded him and other state officials to purchase a local den of iniquity in San Luis, *El Toro Rojo,* the Red Bull Bar (which we renamed *La Vaca Roja*), and transform it into an outpatient clinic and nursing home for those with late-stage Huntington's. We were happy when construction finally began. But it would take ten long years, and lots of fundraising, before the transformation was complete.

Meanwhile new families with *el mal,* some unrelated to the largest kinship with HD in the region, continued to turn up, occasionally in unexpected places. We created pedigrees for them too because people with a distinct ancestry helped the gene search by incorporating different forms of the marker into the mix. Also, we aimed to include all the local families with Huntington's—not only the descendants of María Concepción.

While passing by a flower shop near the Hotel del Lago one day, we noticed that the florist was slightly wobbly. We decided to buy some flowers in order to talk to this florist, who confided to us that *el mal de*

*San Vito* was in his family too. He told us his great-grandfather had been a Spanish aristocrat who had "purchased" a Guajira Indian princess and set her up in a house separate from the one he shared with his wife. He had children with the princess, who eventually developed *el mal*, as did her children and grandchildren.

We said we would like to meet the rest of his family. We had very few people of Guajiro descent with Huntington's in our pedigrees so far and wondered if Guajiro ancestry might confer protective gene variants and, if not, new recombination events.

The florist's family lived in San Rafael de El Moján (typically known as El Moján), a city about twenty-five miles to the north of Maracaibo, on the shore of the lake, and twenty miles from the border with Colombia. There we found them living in poverty, with a mother taking care of everyone. She denied that *el mal* was in her family. I told her that the illness is in my family too, that there's no shame in having it. At first, she didn't believe me but then she burst into tears. She understood. They were the largest family of Guajiro origin we had seen personally. They seemed as vulnerable to the disease as anyone else.

While in El Moján, we discovered another family descended from María Concepción. This family had more education than most, and like many of María Concepción's descendants, they were beautiful, gracious, and warm. The mother was a *bruja buena* or good witch, a healer, which was a tradition in El Moján. She prescribed potions for headaches and other maladies. She had built a brick hut in her backyard where she kept little sculptures of San Benito and other saints who protected against all sorts of evils and temptations. Inside the hut she would smoke a cigar and tell fortunes, including ours. I asked her if I was going to get Huntington's and she said, "No, we need you to find a cure." She was also quite an entrepreneur. In addition to prescribing medicines and telling fortunes, she had a black-market trade going on across the Colombian border. Her husband had died with Huntington's, so she was now caring for all the kids. While we were with her family, people interested in *el mal* would wander in to talk to us. One of the visitors had had juvenile onset while another had a very late onset. We had hoped to find out how the older man explained his good luck but unfortunately we never did.

With everyone we met and whose information we recorded, we had to have absolute certainty about their clinical status because the studies in the lab depended entirely on the accuracy of diagnoses in the field. So sometimes Jim Gusella wanted us to revisit people and ask if they might be willing to donate blood a second time for confirmation.

On one occasion, Jim made a request that took us into new territory, both geographically and, for me, emotionally. Jim wanted us to revisit Sueño, a young woman also of Guajiro ancestry, whom we had met a year or two earlier along with her siblings and other relatives, all of them living in Maracaibo at that time. Jim had found—or thought he found—that Sueño's DNA had a recombination event that could help with isolating the gene. But he wanted confirmation. We discovered that Sueño was now living outside of El Moján, up in the mountains. From our headquarters at the Hotel del Lago, we rented a small bus and went winding up a mountain road to a church, where, we were told, Sueño was a teacher. There we found her among a group of children all dressed in their best clothes. They sang for us in their gorgeous tiny voices, and afterward we explained our mission. Sueño understood and was willing to donate blood again while the children sang songs in their beautiful, high-pitched voices, like little angels. In this magical setting, I remember thinking, or perhaps feeling, that we were certainly going to find a cure.

And then there was serendipity in some of our efforts. We regularly went to a shop in Maracaibo called *Turismo Tropical* to buy typical Venezuelan souvenirs, including hammocks and jewelry. One year, we happened to tell the shop owner we were interested in *el mal de San Vito*. He invited us to come with him to his home in a middle-class neighborhood of downtown Maracaibo. On a bed in the front room lay his wife, with *el mal*. The couple had eight or nine kids and many grandkids. They were an educated family; they read books and even knew a lot about Huntington's. We explained our project to the wife and the shop owner, who agreed to participate, and then we set up a blood-collecting station in the living room where we also did neuropsychological and neurological tests. Afterward they brought us to the house of an uncle whose mother was one of three sisters with super-late onset, in their sixties and seventies, like the uncle himself.

At that point, we were very interested in seeing what elements might be protecting these late-onset individuals, a question we continue to study. The wonderful thing was that everyone we saw was eager to do anything they could to benefit their children and grandchildren if not themselves. That was key for them.

♦ ♦ ♦

One year our team almost didn't come back from a trip to Laguneta. We were returning to Maracaibo after an especially busy day in the village, making the six-hour boat trip on the lake at night this time, not our usual practice, because the winds often started up in the evening. The boat captain told us night travel wouldn't be a problem; he assured us of the safety of the boat—rented from an oil company and smaller than the one that had dropped us several days before. We started off at 6 p.m. under a clear sky, but the water outside the lagoon was already choppy and getting more so as the hours passed. Suddenly an enormous wave came crashing over the bow and smashed out the front windows of the cabin, leaving us all soaking wet and freezing. We asked for life jackets, but there weren't enough for everyone. The crew started pumping out water with a little bilge pump that kept sparking. Everyone started throwing up, except Anne and me. One of our team was so violently ill and despairing that she threatened to jump off the boat; she was convinced that we were all going to die anyway, of dehydration and hypothermia if not of drowning.

I remember asking the man steering the boat if they could retrieve our hammocks from the hold because everyone was freezing; he went off to ask the captain, leaving the wheel unattended! Instinctively I grabbed that wheel remembering what Dad always said on Lake Tahoe about steering on the quarter. In my mind, I was back on Tahoe, steering our motorboat at an angle across angry waves. Finally, the crew member returned with the hammocks and we covered everyone up, pleading with him to radio for another boat to rescue us. "Tell them this is an emergency!" we said. "Tell them we're taking on water, we have no front window, we haven't got enough life preservers." The captain, now up on deck, tried to pretend that he had reached headquarters and they were sending another boat, but my Spanish was good enough by this time to know that they weren't. They'd said the weather was too bad, that we were on our own.

Eventually we saw lights in the distance and thought there was hope, but the captain kept steering past them. He told us they were propane gas rigs and very dangerous; if we tried to land on one and the metal of our boat sparked against the metal rig, we were done for. We implored him to try anyway, and he steered toward a rig. The men on the platform shouted at us to keep away but eventually gave in, ordering us to keep our boat from touching it. Anne and I more or less slung people off the right side of our boat and onto the rig, timing our motions with the rhythm of the waves. We managed to get everyone off, green and miserable, huddled in their underwear on a rig completely unprepared for such visitors. The rig's captain was amazed that our boat crew had ventured out on such a perilous night, saying that this storm was one of the worst of the year. It was a miracle we were alive. He asked what we were doing in Laguneta and I said we were trying to cure *el mal de San Vito*. The captain and all the rig's crew responded that we must be going to find the cure because otherwise God wouldn't have saved our lives.

*Recovering at Canaima, state of Bolivar, Venezuela. (Photo by Anne Young, courtesy of Cambridge University Press. From Young AB. 2025.* Disorderly movements: a neurologist's adventures in the lab and life. *Cambridge University Press, Cambridge, MA.)*

A few days after we returned to Maracaibo, several of us flew to south-eastern Venezuela to see Angel Falls, in Canaima National Park, a UNESCO World Heritage site and our reward to ourselves for surviving the boat trip. We stopped off in the village of Canaima, where Anne took this photo. I liked it so much that she had it enlarged, and I hung it on our living room wall, where it still hangs today.

♦ ♦ ♦

In early 1993, we were once again preparing to leave for Venezuela, packing everything up in my office and beset with all our pre-Venezuela anxieties when Jim Gusella called. A call from Jim portended nothing remarkable. We spoke on a regular basis about the details of the study, checking samples, and so on. We were discussing topics on the agenda for an upcoming event, and Jim said, "But you know, instead of talking to them about that, we could talk about finding the gene."

Nancy: "Well, that would be nice, if we had it."

Jim: "We do."

It was as simple as that. I remember my reaction to the news of finding the marker that brought coworkers running, but my response to the news that the gene had been found isn't etched so indelibly in my brain. I only know it was considerably more subdued. I do remember instantly feeling the responsibility of safeguarding world-shaking news. Jim was given to understatement, but he also had reason to be circumspect. He said to me, "Go down to the end of the hall and stand by the fax machine; I'm going to fax you something, but I don't want anybody else to know. Not even Judy. I'm faxing you a draft of the paper." Jim was worried about being scooped, because of what had happened with the marker back in 1983. This time he was taking no chances.

So, I went down to the end of the hall and stood by the fax machine. Out came the paper. The experience was almost surreal, standing alone in that tiny room on the third floor of the New York State Psychiatric Institute on Haven Avenue with this tremendous secret. It was awkward because I couldn't say anything to Judy or jump up and down or shout hallelujah. I asked Jim if I could tell Dad and Alice and Herb, and he said yes, but no one else. He said he was going to call each member of the collaboration and let them know. One by one, he called them and said,

"Stand by your fax machine!" Later that night, Jim and I had a long phone conversation, and he explained that the disease-causing gene contained an expanded triplet repeat. This meant that unlike the usual gene at that locus, which typically included a limited series of CAGs—nucleotides spelling out instructions for making the amino acid glutamine, a building block of the protein—the gene of individuals with Huntington's had many more of these CAGs. How many was too many? We didn't yet know, but clearly above a certain threshold these extra glutamines distorted the protein made by the gene, turning it toxic to the brain.

I spoke to John Wasmuth, who couldn't believe it, and to David, and others in the collaborative group. I can't recall whether this was a waking dream or a real dream, but I kept seeing the families in Venezuela with CAGs written all over their bodies. But in my mind's eye, instead of harming them, the letters formed a beautiful design.

I recalled a meeting of the Gene Hunters soon after the Fragile X discovery, when I got out the Venezuela pedigree and looked at one of the big families whom I had gotten to know well and to whom I was very attached. There were nine children in this family and seven of the nine got sick, some starting in early childhood and others in their teens or early twenties. Several families resembled this one, with parents whose symptoms started in their thirties or forties while their kids with Huntington's got sick at much younger ages. I recalled looking at the pedigree and saying, "Look, this is anticipation, we have anticipation! We must have the same kind of expansion as Fragile X!" And David Housman replying, "Absolutely, definitely, Huntington's is a disease that has expanded repeats. That's what's causing it, a huge expansion." We were very excited, and everyone was saying we just needed to look for an expansion. That was a physical landmark we could search for.

Jim's group had located a promising candidate region, and David's group decided to scan it for expansions. But, thinking of the large expansions in Fragile X, they set a lower limit around one hundred repeats, meaning below that threshold an expansion wouldn't be visible. It was a bit like being on the surface of the ocean and scanning with sonar for hidden icebergs above a certain size. Smaller icebergs, below that threshold, wouldn't register. They would be invisible. So, David didn't get any hits. His result was dismaying because we felt confident that this region contained the Huntington's gene, and we were excited that an expansion would explain clearly what we were seeing clinically. We were all disappointed once again.

At that point, Jim and his team were trying to narrow the candidate region while also looking for genes in the neighorhood that varied among individuals and might be useful as markers. Fast forward two years. They came upon a gene with forty-eight repeats, considerably more than the number of repeats they had seen on any non-HD chromosome, but fewer than in Fragile X and myotonic dystrophy. They decided to look again at the DNA of the little boy who got sick at two years old—the boy I had met on my first visit to Laguneta in 1979. They wanted to see what his DNA looked like at that spot. They had never really been able to see his DNA properly on the gels that they used to visualize DNA because the fragments always balled up into a big smudge that they dismissed as an artefact. But this time they could see it more clearly. They also looked at his other close relatives.

And there it was. The father with HD, whose symptoms started in his early forties, had slightly fewer than forty-eight CAG repeats; the children who became symptomatic from twenty-six to eleven years old had CAG sequences of increasing size the younger the age of onset. The CAG length of the mother without HD was around twenty. And the DNA of the little boy who developed symptoms at the age of two appeared to have close to one hundred CAG repeats.

But the story was not yet over. This was just one family. They needed to look at many more. Jim and his team ended up studying a total of seventy-four more families with Huntington's, all of whom showed the same pattern: a longer series of CAGs on the chromosomes of those with Huntington's than those without. It was clear that this gene had to be an expansion.

Jim says that it was the consistency across all seventy-five families that provided the conclusive evidence. But I prefer to think it was an angelic little boy with blue-green eyes and freckles, from Laguneta, in the state of Zulia, Venezuela, who led the way in solving the mystery of the long-sought-after gene causing Huntington's disease.

♦ ♦ ♦

A few days after Jim's phone call, Judy and I flew to Venezuela for our annual visit. Keeping the secret of the gene discovery was one of the hardest challenges of my entire life. I could barely keep myself from bursting out with the news. But I knew any revelation would endanger our publication and that all the good feelings we had built up within the collaboration over the previous decade would go out the door. The paper was going through yet more drafts, which Jim now faxed to the Hotel del Lago where the fax machine didn't work well. I was trying not to let Judy know that anything unusual was going on, although I kept running downstairs to the front desk to retrieve faxes. Jim worried someone in Venezuela might read the faxes, but I told him I was getting them myself, so no one else was looking at them. I did worry, though, that the identities of some of the Venezuelan families would be obvious in the paper, and there were other things we were trying to change in the text, going back and forth with the faxes, arguing even about who to name as members of the collaboration and how to write the acknowledgements because we hadn't really worked out all the details.

Jim wanted to have a press conference at Harvard, and I wasn't going to miss this one. We both agreed that the announcement must be made in a way that reflected the collaborative process, well established after so many years of working together.

Within a few days, *Cell,* a prestigious scientific journal, accepted the paper, although it would still take some time before it went to print. But now we could let the others know. I asked Judy to come for a walk with me at night, and we went out on the little pier behind the Hotel del Lago, which faced directly onto the lake. A high wind was blowing, and big white-capped waves were slapping at the pier. And then I told her.

"What?" she said. She was totally bowled over. "Yes," I replied, "we found the gene." She couldn't believe it. "Are you sure? Are you really sure?" she asked. "Yes, we really found the gene, and now I have to go back to Boston to the press conference."

At that point, except for Anne (who had succeeded Joe Martin as Chair of Neurology at Harvard Medical School and chief of neurology at Mass General), none of the other team members who were with us in Venezuela knew. We decided to have a pool day the next day. Occasionally we would have a pool day to work on the data, restock bags with medicines and forms, and have a chance to relax. While everyone was hanging out by the hotel pool, I asked Marshall Fordyce, then just a freshman in college and member of our team that year, if he could film us because we had no footage of the team in this casual setting. I was very fond of Marshall, who had a gift for entertaining the children while we did our work. (He eventually went to Harvard Medical School and subsequently trained in infectious diseases at Columbia, going on to found several biotech companies. He also happened to be the grand-nephew of Herb's and my friend, the great philanthropist Mary Lasker.)

As Marshall started to film, I stood up to make the announcement so everyone could hear. "Well," I said. "I just want to give you a little news. We found the gene."

Pandemonium! Everyone was jumping up and down, hugging and kissing each other, and shouting congratulations in Spanish and English while Marshall was trying to film.

The following day, Anne and I flew back to Boston for a week to attend the press conference, not at the Ether Dome this time but elsewhere at Massachusetts General Hospital. Most of the PIs from the Gene Hunters came and each one of them spoke to the overflowing crowd. The discovery got a huge amount of media attention, because it was the first time that a DNA marker had been used to find a disease gene that could have been anywhere in the genome. *The New York Times'* morning edition ran it on the top of the front page, as did the *Los Angeles Times*, which quoted the director of the NINDS saying the discovery "may prove to be the crown jewel of recent neurogenetic discoveries."[1]

That night, we had a big celebration party in Boston. Herb came up from New York despite a huge blizzard and everyone was overjoyed.

Then Anne and I returned to Maracaibo, where Judy Lorimer and Bob Snodgrass, a neurologist from Los Angeles Children's Hospital, had kept the project moving beautifully. We decided to have a party for all the families. We chose *La Vaca Roja*—which was already in the process of transforming into a nursing home—and Margot and I got food, partly donated and partly paid for by the government. We rented music for the afternoon. We wanted to celebrate and thank everyone for their participation over many years and their immense contribution to the research.

Cell, Vol. 72, 971–983, March 26, 1993, Copyright © 1993 by Cell Press

# A Novel Gene Containing a Trinucleotide Repeat That Is Expanded and Unstable on Huntington's Disease Chromosomes

**The Huntington's Disease Collaborative Research Group***

**Summary**

**The Huntington's disease (HD) gene has been mapped in 4p16.3 but has eluded identification. We have used haplotype analysis of linkage disequilibrium to spotlight a small segment of 4p16.3 as the likely location of the defect. A new gene, IT15, isolated using cloned trapped exons from the target area contains a polymorphic trinucleotide repeat that is expanded and unstable on *HD* chromosomes. A $(CAG)_n$ repeat longer than the normal range was observed on *HD* chromosomes from all 75 disease families examined, comprising a variety of ethnic backgrounds and 4p16.3 haplotypes. The $(CAG)_n$ repeat appears to be located within the coding sequence of a predicted ~348 kd protein that is widely expressed but unrelated to any known gene. Thus, the *HD* mutation involves an unstable DNA segment, similar to those described in fragile X syndrome, spino-bulbar muscular atrophy, and myotonic dystrophy, acting in the context of a novel 4p16.3 gene to produce a dominant phenotype.**

**Introduction**

Huntington's disease (HD) is a progressive neurodegenerative disorder characterized by motor disturbance, cognitive loss, and psychiatric manifestations (Martin and Gusella, 1986). It is inherited in an autosomal dominant fashion and affects ~1 in 10,000 individuals in most populations of European origin (Harper et al., 1991). The hallmark of HD is a distinctive choreic movement disorder that typically has a subtle, insidious onset in the fourth to fifth decade of life and gradually worsens over a course of 10 to 20 years until death. Occasionally, HD is expressed in juveniles, typically manifesting with more severe symptoms including rigidity and a more rapid course. Juvenile onset of HD is associated with a preponderance of paternal transmission of the disease allele. The neuropathology of HD also displays a distinctive pattern, with selective loss of neurons that is most severe in the caudate and putamen. The biochemical basis for neuronal death in HD has not yet been explained, and there is consequently no treatment effective in delaying or preventing the onset and progression of this devastating disorder.

The genetic defect causing HD was assigned to chromosome 4 in 1983 in one of the first successful linkage analyses using polymorphic DNA markers in humans (Gusella

***The Huntington's Disease Collaborative Research Group comprises:**

***Group 1:***
**Marcy E. MacDonald,[1] Christine M. Ambrose,[1] Mabel P. Duyao,[1] Richard H. Myers,[2] Carol Lin,[1] Lakshmi Srinidhi,[1] Glenn Barnes,[1] Sherryl A. Taylor,[1] Marianne James,[1] Nicolet Groot,[1] Heather MacFarlane,[1] Barbara Jenkins,[1] Mary Anne Anderson,[1] Nancy S. Wexler,[3] and James F. Gusella[1†]**
[1]Molecular Neurogenetics Unit
Massachusetts General Hospital
and Department of Genetics
Harvard Medical School
Boston, Massachusetts 02114
[2]Department of Neurology
Boston University Medical School
Boston, Massachusetts 02118
[3]Hereditary Disease Foundation
1427 7th Street, Suite 2
Santa Monica, California 90401

***Group 2:***
**Gillian P. Bates, Sarah Baxendale, Holger Hummerich, Susan Kirby, Mike North, Sandra Youngman, Richard Mott, Gunther Zehetner, Zdenek Sedlacek, Annemarie Poustka, Anna-Maria Frischauf, and Hans Lehrach**
Genome Analysis Laboratory
Imperial Cancer Research Fund
Lincoln's Inn Fields
London, WC2A 3PX, England

***Group 3:***
**Alan J. Buckler,[1] Deanna Church,[1] Lynn Doucette-Stamm,[1] Michael C. O'Donovan,[1] Laura Riba-Ramirez,[1] Manish Shah,[1] Vincent P. Stanton,[1] Scott A. Strobel,[2] Karen M. Draths,[2] Jennifer L. Wales,[2] Peter Dervan,[2] and David E. Housman[1]**
[1]Center for Cancer Research
Massachusetts Institute of Technology
Cambridge, Massachusetts 02139
[2]Division of Chemistry and Chemical Engineering
California Institute of Technology
Pasadena, California 91125

***Group 4:***
**Michael Altherr, Rita Shiang, Leslie Thompson, Thomas Fielder, and John J. Wasmuth**
Department of Biological Chemistry
University of California
Irvine, California 92717

***Group 5:***
**Danilo Tagle, John Valdes, Lawrence Elmer, Marc Allard, Lucio Castilla, Manju Swaroop, Kris Blanchard, and Francis S. Collins**
Department of Internal Medicine and Human Genetics
and The Howard Hughes Medical Institute
University of Michigan
Ann Arbor, Michigan 48109

***Group 6:***
**Russell Snell, Tracey Holloway, Kathleen Gillespie, Nicole Datson, Duncan Shaw, and Peter S. Harper**
Institute of Medical Genetics
University of Wales College of Medicine
Cardiff, CF4 4XN, Wales

*†Correspondence should be addressed to James F. Gusella*

Cell, *March 26,1993.*

And we also had to figure out what to tell them. We needed to make clear that although it wasn't the cure, it was a huge advance. We had found the main cause of the disease and were closer to understanding it. We didn't want to give them false hope, but at the same time we felt it was important to celebrate along the way. Our explanation didn't entirely satisfy them, but they weren't going to turn their backs on a party. In fact, we had two parties, one in San Luis and one in Barranquitas, where the people from Laguneta also came. The kids were dancing the "dirty dog" and everyone else was dancing up a storm. Even Negrette came, celebrating with us. Our team members were the servers, including Anne, future president of the Society for Neuroscience and the American Neurological Association, who stood with me in the serving line handing out food.

♦ ♦ ♦

The Gene Hunters didn't immediately disband. We even went to Islamorada the following May where Dennis gave everyone gifts for finding the gene, now called the *huntingtin* gene. We didn't know exactly what to do next. I remember one day, not long after the gene discovery, walking with David Housman through Central Park as he pointed out the wildlife in the middle of Manhattan. We agreed that we had to go back to the proactive outreach we had followed to find the Gene Hunters. We decided to get our advisory committee together and ask them where we should go from here. What should be our next priorities? Their responses to us were these: We need animal models. We need to explore possible modifiers. We need to understand the huntingtin protein and what it does in its normal and aberrant states. We made a list of all the tasks that needed to be done and resolved to go out and find the best people who could do them. There were some hurt feelings and tensions along the way, and complaints about too many meetings, and not wanting to share, and so on. It wasn't always wonderful.

Meanwhile our annual visits to Venezuela continued. One of the benefits of returning year after year was getting to know the families over time, trying to help them, and seeing how they managed the disease through the years. By observing and testing them when they were willing (and most were)—cognitively, neurologically, and psychologically—we could follow in detail their progression and try to see where a therapy might intervene most effectively. We could observe the first subtle signs of onset and offer aid when possible (not only for *el mal,* of course, but also for all the common maladies that added to the misery of the families). We learned to appreciate their way of conceptualizing *el mal.* They believed that everyone whose parent is affected inherits the disease but only some get sick; this way of thinking astutely acknowledges the weight of the disease on *all* members of the family and not only on those who develop symptoms. The Venezuelans we met were also expert diagnosticians—especially Margot, who could distinguish between those whose difficult behavior was an aspect of the disease and those who might act *loco* but didn't have *el mal.*

When we first began working in Venezuela, we had been impressed with the sight of people with severe chorea walking in the streets of San Luis and Barranquitas, out and about in the local community, seemingly integrated and accepted, although they confided to us that they were sometimes bullied or mocked. But soon we realized that those in the late stages were often confined at home, left in a hammock or on a bare concrete floor while their relatives went out to work. Food in these communities could be in short supply, for the healthy as well as the sick and disabled, and people with Huntington's need many more calories than usual for their age (on account of the movements that use up calories, and possibly also due to metabolic differences that we don't fully understand). These individuals might be more or less abandoned or at least left alone for long periods of time. But where else could they go?

So we were thrilled when Margot's plan for a nursing home and clinic finally came to fruition in June of 1999 when the Casa Hogar Corea de Huntington Amor y Fe opened to great fanfare even as workmen were still completing construction. Where the raunchy old bar had been stood a new two-story building with garden patios graced with small trees and flowering plants. I was astonished that so much work had gone on since my previous visit a few months earlier, thanks to Margot and to the governor's wife who had taken a personal interest in this project.

On opening day, we marveled at the light-filled interior containing spacious rooms for the patients, patios, consulting offices for clinicians, a dining room, reception area, laundry, a kitchen with an industrial size stove and refrigerators, and a dayroom with a television among other amenities.

We had decided that the staff should be Huntington's family members. We would draw on the skills that many already had for compassionate care while also giving income to desperately poor families. As director of the Casa Hogar, Margot offered additional training to

*The Casa Hogar Chorea de Huntington Amor y Fe (Huntington's Chorea Home of Love and Faith) nursing home/hospice in San Luis. Courtesy of HDF.*

a core staff of about thirty, while also employing two highly trained nurses. (I paid part of the expenses of the Casa Hogar out of my NINDS Venezuela grants and later grants from the W.M. Keck Foundation, while a wealthy Venezuelan family provided generous support for a number of years, as did the Hereditary Disease Foundation and Anne Young and Jack Penney through their HD Venezuela Family Foundation.) We were able to keep the lights on for more than a decade, along with staffing a clinic at the Casa Hogar for Huntington's outpatients who came every day, around seventy per week. And we provided meals to many more.

I was proud of the Casa Hogar and grateful for Margot's loving dedication and tender care for these patients. Margot understood them and knew how to train others to care for them. To be able to give people who had suffered so much this safe harbor, this haven where they would be cared for with dignity and respect—that meant everything to me.

Over the years, the Casa Hogar grew increasingly crowded, however, so great was the need in the barrio for its services. Originally designed for sixteen patients, it soon housed thirty-five. And more kept coming. In the early 2000s we received added support from an unexpected source, an eminent British pharmacologist, Michael Rawlins—later "Sir" Michael Rawlins—who merits special acknowledgement for his wide-ranging efforts relating to the Casa Hogar and Huntington's research generally.

I had met Michael, then a professor at Newcastle University in the United Kingdom, at a memorable meeting on public health policy at the nonprofit research organization, RAND, in Santa Monica—memorable in part because the ceiling of the room partially collapsed during the meeting, sending us fleeing outside. At RAND he heard about Huntington's from me and would subsequently do important research on its incidence and prevalence worldwide and in the United Kingdom specifically. Michael attended many HDF meetings where his expertise in statistics and clinical trials, along with his wit and generosity of spirit, illuminated a wide range of discussions. (He would later become the chief architect and first chair of NICE, the National Institute for Clinical Excellence—now the National Institute for Health and Care Excellence, a body that makes recommendations about drugs and medical devices to be covered by the National Health Service.)

When I told him about the crowding at the Casa Hogar, he offered to help, rallying invaluable support from his friend, the British ambassador to Venezuela at that time, Catherine Royle. Together they managed to secure resources so that Casa could be expanded, including installation of a new up-to-date kitchen, to accommodate the greater number of patients, as many as sixty-five in the final years.

*Susannah Rawlins, me, Michael Rawlins, and Carl.*

Unfortunately, the attempted coup against Hugo Chávez in 2002 ended our annual study expeditions. Washington advised us not to return to Venezuela, although I did make several subsequent visits on my own, while Michael and Ambassador Royle, who became a friend, also did what they could to help.

With conditions deteriorating during Chavez's final years, Margot moved to Texas where one of her daughters, a neurologist, lived. For several years Margot oversaw the Casa Hogar remotely while the nurses in Venezuela continued their work for as long as they were able. But by about 2014, donations had dried up and we were no longer able to pay salaries or buy medications and supplies to keep the Casa Hogar going, although it maintained some outpatient services for a while.

During this same period, we also sought help for Barranquitas, where the housing and sanitation situations were even more dire than in San Luis. A ditch running through the community and serving mainly as a sewer was the only source of "running" water in the area. Through Margot I had met Donal O'Neill, an Irish engineer and Shell Oil employee who had come to Maracaibo in 1996 to oversee Shell's "reentry into Venezuela." Shell was an important employer in the region, and once we introduced Donal to the problems of Barranquitas, he got interested in trying to persuade Shell—which had oil rigs nearby in the lake—to support improvements in local housing and sanitation. At the same time, I approached Frank Gehry who suggested contacting the Southern California Institute of Architecture, better known as Sci-Arc, where students might be recruited to come up with a proposal for low-cost housing. We managed to interest a number of them in the project. They developed an imaginative and pragmatic proposal for housing specifically attuned to both the tropical climate of Barranquitas and the needs of people living with Huntington's. Unfortunately, Donal wasn't able to persuade Shell to support more than minor improvements in sanitation, and in 2000 he was reassigned to the Netherlands. The project, sadly, has languished.

CHAPTER

# 9

# MAPS, MICE, AND MODIFIERS

From the first moment I heard about it, sometime in the mid-1980s, the idea of mapping all human genes captured my imagination. The prospect of knowing ourselves on this intimate level enthralled me and made me want to be part of the project. I especially liked its egalitarian dimension. Mapping all our genes could benefit research not only on the big "have" diseases (those like Alzheimer's with considerable federal funding) but also the smaller "have not" diseases like Huntington's (with little federal funding). In fact, our identification of a genetic marker for Huntington's gave a big boost to the momentum for the Human Genome Project, which got under way in 1990.

Two years earlier, I had been invited to a meeting in Valencia, Spain, to speak about the medical impact of mapping all human genes. At the time this topic wasn't central to the genome discussions, which were focused more on cost and scientific strategy. In Valencia I met James Watson, co-discoverer of DNA's double-helix structure, who was about to become director of the National Center for Human Genome Research at the NIH. The NIH director, Jim Wyngarden, was also at the meeting, as were Victor McKusick, known as the "father" of medical genetics, Norton Zinder from Rockefeller University, and Charles Cantor, a colleague at Columbia whom I knew from the early days of the Gene Hunters. These and other leading lights of human genetics and molecular biology were there—and me, a clinical psychologist, an associate professor, a person at risk for Huntington's, and a woman, whom they nonetheless treated as an equal. They were looking for people to serve on an advisory committee to the genome project, and they must have realized they didn't have any women. They probably recognized that with me they could have a social scientist, a woman, and a consumer of genetic services altogether; they could fill three slots for the price of one. Plus, I was working closely with geneticists and understood the science. Whatever their motives, I was thrilled to be a part of it.

As I recall now, Jim Watson was the one who invited me to join the advisory committee. Nearly everyone on that committee also became chair of a genome working group. I suspect that because I had done research on the psychosocial ramifications of genetic risk, had been involved with genetic testing and counseling, and had personal experience living at high genetic risk, Watson steered me to ELSI—the Ethical, Legal, and Social Issues Working Group.[1] Jim understood from the start that any project in genetics needed to address the social dimensions, because genetics had long been tainted by its historical association with eugenics. I was someone directly affected by that social legacy. And I was deeply drawn to the idea of science being introspective about itself. I wasn't an ethicist, but the combination of my research and lived experience was something unique that I could bring to the table. I was grateful to Watson for this opportunity. He and his wife Liz became good friends, and I found him very supportive of women in science. I was also moved by his devotion to his son with schizophrenia, who lived with them and was a focus of great concern.

*With Jim Watson (*left*) celebrating his birthday.*

At Watson's insistence, ELSI became the first federal science initiative to allocate "science" money for social science research, between 3% and 5% of the entire genome budget. It was obvious that the new stores of genetic information coming out of the Human Genome Project would have a huge impact on such issues as prenatal testing, privacy, disability, and discrimination in insurance and employment, among other arenas. But ELSI was controversial from the start, even within the NIH. Some researchers argued that its expenditures were wasted money that could have been spent on "real" science. Others warned that public discussions of ethical and social issues would stir up opposition, create obstacles for researchers, and impede the progress of science. Speaking from a different vantage point, a third group of critics feared that ELSI was mainly a preemptive strike to thwart criticism and that we were going to be captives of the "genome apologists."

I felt it was useful to have conversations now about what the new technologies could and could not do and to push right away for protections. I wanted to focus on concrete, reality-oriented practical issues, such as improving access to genetic services, developing educational materials, and training more counselors and genetics personnel. I especially wanted to emphasize the impact on patient care. Ultimately, I didn't hear many complaints other than that perhaps we spent too much money. I felt that ELSI helped articulate the values that should govern scientific research and made such discussions ongoing and visible. Whatever its limitations, ELSI helped legitimize this research as part of "science" and secured support for pilot programs in the future.[2]

Among our many activities, we funded conferences and research on such topics as genetic discrimination in insurance, employment, education, and clinical care; genetic privacy; prenatal and predictive genetic testing; and ways to ensure benefit from these tests. One of our working group's first successes, in fact, was to move the NIH toward supporting pilot projects to study genetic testing for cystic fibrosis, the most common genetic disease among Caucasians. After that, other institutes within the NIH followed suit. I felt I could channel my triple identity as a researcher, clinician, and person at high risk for a neurodegenerative disease to help frame discussions and make policy recommendations. I never forgot that I had a personal stake in the issues. The questions being discussed were not abstract for me. I wanted to emphasize that whatever challenges and difficulties the vast new fund of genetic information might produce, we could not allow our concerns to override the need for science to advance its efforts to alleviate the suffering caused by hereditary disease.

Of course, ELSI was also dealing with huge social and political problems far beyond our purview, such as those resulting from our fragmented health insurance and long-term care systems. But we could try to catalyze a political constituency to start implementing policy changes that ELSI recommended. We had to walk a fine line between being activist and pushing for change on the one hand and, on the other, being held accountable for all the failures that we knew were going to come.

The Americans with Disabilities Act (ADA), which ELSI supported, marked one important step toward protecting those with genetically based disabilities against discrimination as it related to employment. As a witness at the Senate hearings, I observed Senators Al Gore and Ted Kennedy as Jim Watson presented our case. At one point, Jim quoted a doctor's pejorative description of a woman as "feebleminded," and Senator Kennedy became visibly disturbed. I imagined he was remembering how his sister Rosemary's potential had been decimated by a lobotomy. At that moment, I knew we had gained a supporter. The ADA, passed in 1990, was an historic landmark, "the most important piece of human rights legislation since the Civil Rights Act" of 1964, in the words of Senator Kennedy.

However, except for limited protection related to jobs, legal protection for those with genetic risk for *future* disability didn't come until much later. In testimony about ELSI before a Senate subcommittee in October of 1991, I laid out in personal terms some of the issues for people like me. I spoke about the right to privacy. But I also talked about the "essential rights of persons to *disclose* genetic and medical information with impunity, without fear of harmful reprisals." I described learning about Huntington's disease in my family at the age of 23 and then going immediately to graduate school, where at first I was "embarrassed and ashamed to tell anyone about my mother's sad decline or my own risk." I was afraid people would treat me differently, watch me for symptoms, not want to date me, or be overly distant or too solicitous. I told how, in graduate school, I began working with families with Huntington's disease, but for some time kept that world and my academic life quite separate; how it felt slightly schizophrenic, literally commuting between my world of HD families in Detroit and my academic life in Ann Arbor, Michigan.

I spoke of how I asked all my advisors "to rewrite their letters of recommendation because, although they had said nothing explicitly, I was afraid that it would be too obvious why I was interested in Huntington's disease. I was very concerned that I would never be hired if my risk status was known, and certainly never be considered for tenure." I told how the turning point came for me when the NINDS invited me to chair the Congressional Commission on Huntington's and later join their staff. "For the first time, I could be totally open about the disease without fear of alienating my colleagues or losing my job security and employment benefits." And I described how being offered a civil service position at the NINDS, "from which it is difficult to extract people," was "enormously healing, because my colleagues, experts in the disease, were willing to take a chance on a person with a one in two possibility of developing a neurodegenerative disease of the brain and body, with insidious onset, causing failures of judgment and memory and emotional instability. It was not privacy in this instance that was necessary—it was candor."

In speaking about the benefits of the Human Genome Project for those dealing with hereditary diseases but also with diseases such as cancer and Alzheimer's that were most often sporadic rather than familial, I stressed that genes were still often involved and that new therapies might intervene in damaged genes, whether that damage was inherited or caused by environmental toxins. I noted that "it would be a bitter irony if people who can benefit from early diagnostic [genetic] tests are dissuaded from availing themselves of the test because they may lose the very insurance they need to prevent the disease...". I also noted how genetic information about one individual may immediately reveal information about others in the family. "When I took my mother to a new doctor one day, he said to me, 'Oh, Huntington's—you have a one in two chance of having it too, no?' He had no idea what I knew or didn't know." I talked about the complexity of many genetic conditions, which vary tremendously in ways that presymptomatic tests may not be able to capture. I asked the representatives, "If you knew for certain that I was going to develop Huntington's disease, would you feel any differently toward me now?" And I made a pitch for the necessity of advancing genetics research despite all the complexities. "The fruits of the Human Genome Project are a source of great hope for millions of Americans," I told the committee. "When I ask people who are presymptomatically diagnosed with Huntington's disease what sustains them, they usually answer 'God and trust in science.' Our political agenda is complex and will demand empathy, caution, and courage. It cannot be carried out at the expense of science, but as two complementary programs to prepare for the twenty-first century."[3]

That has been my philosophy all along, but the road to the Genetic Information Nondiscrimination Act, or GINA, which ELSI also supported, was long and arduous. This law specifically outlaws discrimination based on genetic status in health insurance and housing as well as employment. First introduced into the House and the Senate in 1995, the bill didn't pass until 2008, when President George W. Bush finally signed it into law.

♦ ♦ ♦

The discovery of the *huntingtin* gene opened up an array of questions about the meaning of the expanded sequence of CAG repeats and whether the gene in different tissues could have different size expansions. Usually, the mistake is the same in all tissues; we had seen identical size expansions in blood, skin, muscle, and brain (from autopsies). But what if the size expansions in the germ line, the sex cells that pass the mutation from one generation to the next, differed from those in somatic cells such as blood and skin?

One day in1994, as our team was about to leave for Maracaibo, Jim Gusella called me with an unusual request: "Do you think it's possible to get semen samples in Venezuela?" He wanted to see how stable the DNA was in sperm. We knew by now that most of the children with juvenile Huntington's had inherited it from fathers who developed symptoms as adults. What do fathers have that mothers don't? Sperm. But how to collect it! I was dubious about accomplishing this request. We had tried in the States and got very few donors. Either men didn't want to participate or they couldn't. Nevertheless, we agreed to try. But how? I didn't even know how to ask. We understood enough about the culture to know that for some men this would be embarrassing and potentially humiliating; we would be asking them to masturbate, which was considered taboo. So first, Margot and I tried to learn the local vernacular without revealing why we wanted to know it. We also wanted to be clear that we weren't asking them for urine samples.

Then one day, in Barranquitas, we found ourselves in a sweltering bare cinder-block room in a tiny medical clinic, with several young men facing us, seated on boxes or folding chairs. We had to close the windows against the curious children peeking through, and we were all sweating profusely. But elegant and very proper Margot lent a crucial dignity to the situation. She explained that with the discovery of the gene, we needed to understand what happened when this gene was passed from parents to children and especially from fathers, who were much more likely than mothers to have young children with the disease. The Venezuelans themselves had noticed that most youngsters who developed symptoms had fathers with Huntington's, so they, too, were curious about the reason why.

Margot did the heavy lifting since this delicate situation really called for a native speaker. She asked what they called "semen" in Barranquitas—naturaleza was the answer—and then she asked if they would be willing to help us understand more about the disease by making a donation into a little tube. We held up a copy of Playboy and pointed toward the clinic's bathroom. My Spanish took a quantum leap forward when I learned the Spanish words for "Come in this tube and close it tight."

There were a few misunderstandings along the way. Some men asked if they could do it at home with their girlfriends or if they could do it with one of us. I think our awkwardness worked to our advantage, calling on their machismo and natural graciousness to put us at ease. And eventually they took pity on us and agreed, although initially most of them preferred to take the collection tubes home. They asked for paper bags so they wouldn't be embarrassed bringing their samples back to the clinic. And they did bring them back. With time, they grew so comfortable with us that we could provide a chair in the bathroom near where we were doing examinations, prop up Miss February on the toilet, and ask them to call us when they were done. Even some men with advanced HD who could barely walk were able to participate. We collected tubes every day. Eventually we collected about 1700 semen samples, with some men providing them year after year.

These samples proved to be extremely important because we discovered a great variation in the number of repeats among sperm from the same individual. Some sperm had the same number of CAGs as in the blood (repeats in the blood cells themselves typically remained uniform). But other sperm from that same man might carry many more CAG repeats, as many as double or triple the number in their blood. So here was the first plausible explanation for why so many of the symptomatic children had paternal inheritance: Fathers often transmitted to their children a much higher CAG number than they themselves possessed. It became clear that one reason why the offspring of men with Huntington's were more likely to become symptomatic at a young age was the greater likelihood of large CAG expansions in sperm.

♦ ♦ ♦

Meanwhile, creating an animal model of Huntington's ranked high on our list of priorities. Now that we had the gene, Gillian Bates, the talented former postdoc with Hans Lehrach, took on the task of creating a mouse with Huntington's disease.

Recall that, as one of the Gene Hunters, Gill in 1989 had cloned the tip of Chromosome 4. It was she who discovered that the tip from individuals with and without Huntington's looked identical. This finding was a major discovery that helped focus the search away from the tip toward a slightly more internal region of the chromosome where the gene was eventually found. She went on to establish her own lab, soon landing at Guy's Hospital at King's College London and subsequently at University College London where her work eventually earned her election to the prestigious Royal Society. Her first order of business after the gene discovery was to try to put the gene into a mouse.

Gill began by going to Cold Spring Harbor Laboratory to take a "make-a-mouse" course. Cold Spring Harbor was at the epicenter of molecular genetics research and at that time was home to eight Nobel Prize winners in physiology or medicine. They ran a three-week immersive course

*(Left to right) Elena Cattaneo, Reiner Kuhn, and Gillian Bates at an HDF workshop.*

in creating mouse models, and Gill came back to London ready to take on this challenging task. In brief, she made a "library" of DNA fragments from a child with an extremely long CAG repeat and took out just the start of the *huntingtin* gene with about 150 CAGs. She then injected this DNA fragment into a mouse embryo and put the embryo into another mouse that gave birth a few weeks later. She repeated this process many times over, with additional mice and DNA from other (human) individuals with different length CAG expansions.

And so was born the R6/2, the first transgenic HD mice—meaning that they contained foreign genetic material (human CAGs). The mice soon exhibited symptoms seen in humans with Huntington's: the unsteady walk, the loss of weight despite eating voraciously, and cognitive decline as they repeatedly pet their noses and groomed their fur. They lived between sixteen days and twenty-one weeks, their lifespan and severity of HD symptoms varying in accordance with the number of CAG repeats they received. The R6/2 mouse marked a major milestone in our ability to explore disease mechanisms and test interventions, so much so that *Cell* made Gill's mouse the cover story of its November 1, 1996 issue, which featured a full-color photographic portrait of the historic yet unassuming R6/2.[4]

After Gill produced her mice, Stephen Davies, a neuropathologist colleague at University College London, decided to follow up on her work by examining these R6/2 mice through an electron microscope, which offers much greater magnification and resolution than a light microscope. He found that their mouse brains had clumps or aggregates of the huntingtin protein, something that supposedly had not been seen in the brains of humans with Huntington's who had died. Around the same time, a neuropathologist at Harvard, Marion DiFiglia, showed that, lo and behold, these clumps did indeed show up in humans. Marion used material from autopsies and pinpointed these aggregates (sometimes called nuclear inclusions) with startling clarity.

Stephen then decided to look through the electron microscopy literature on Huntington's. He found an eighteen-year-old study in which the same clumps were reported in brain biopsy tissue from humans living with Huntington's.[5]

Sometime later, I asked David Housman about the human brain tissue from those earlier researchers. If we'd looked at that time and purified those clumps, I asked David, could we have found the huntingtin protein twenty years earlier? He said yes.

♦ ♦ ♦

A few years after Gill's R6/2 mouse came on the scene, several new developments brought our foundation to a whole new level of activity. In early 1996 we received an extremely generous infusion of funding from an anonymous donor that enabled us to greatly increase our grant giving. Suddenly we had resources to carry out work in a much more expansive way than ever before. The science board decided to form a Cure HD Initiative, or CHDI (Cure Committee for short), within the foundation to speed up the pace of research toward a therapy, hiring Ethan Signer, from MIT, as the full-time executive director. In line with our efforts to be proactive in seeking out promising projects and investigators, the committee began developing contracts to accomplish specific high-priority research aims.

Several years later, with the departure of Allan Tobin, we hired a new scientific director, Carl Johnson, a tall strapping Midwesterner living in Black Earth, Wisconsin, outside of Madison, who had done basic

*Carl Johnson with Bob Horvitz. (Photo by Jeff Szmulewicz.)*

research on the worm *Caenorhabditis elegans,* including on the inhibitory neurotransmitter GABA. After a decade in academia, Carl had worked in biotech. He founded a company, NemaPharm, hoping to use genetically modified *C. elegans* to screen drugs as possible treatments for genetic diseases, an effort that led to his friendship with our collaborator, Bob Horvitz. Eventually Carl moved to the nonprofit world and joined the HDF, bringing with him his experience as an experimentalist, even temper, and leadership and networking skills. I was very fond of Carl, whose warmth and sense of humor made him good company. With Carl at the helm for more than a decade, we were well situated to explore further the implications of finding the *huntingtin* gene.

◆ ◆ ◆

Meanwhile, Gill's work had shown that if you put even a fragment of the mutant *huntingtin* gene into a mouse, the mouse gets sick and develops aggregates in the brain commensurate with the human disease. The R6/2 mouse became such a popular model and so in demand that soon Gill found herself overwhelmed with requests to share it. She couldn't breed mice fast enough.

I went to London to visit Gill shortly after she developed the mice. I held some of them, amazed that these cute little creatures wiggling around in my hand had the same disease Mom had had. They were the very first transgenic animal model of Huntington's. From Mom to mouse model. I found the experience oddly moving, as if physically touching these mice with Huntington's was touching something intimately related to me. Or perhaps they reminded me of my mouse experiment in high school, when the traumatized mice who received affection recovered while the others—at least some of them—died in my hand. The idea that I was a mouse murderer had never entirely left me. Now I had the opportunity to bestow affection on *all* these Huntington's mice, whom I hoped were going to lead the way to a treatment or even a cure.

♦ ♦ ♦

After the R6/2 mouse, researchers developed an alphabet soup of HD mice to model other aspects of the disease. Each mouse model carries a different part of the mutant human *huntingtin* gene and shows distinct symptoms and different patterns of affecting the brain: the N171-82Q, the HDHQ111, the CAG140, the Hdh (CAG)150, the YAC 128, the BACHD. This last one, developed by X. William Yang and his postdoc Michelle Gray at UCLA (now professor at the University of Alabama, Birmingham), became one of the most widely used models after the R6/2. I also held some of these BACHD mice in my hand on a visit to William's lab at UCLA, experiencing the same uncanny sense that I had with the R6/2 of touching a creature that would one day deliver the cure.

During his MD/PhD training at Rockefeller University in the lab of Nathaniel Heintz, William—who received his MD from Weill Medical College of Cornell University—had developed a new way to generate transgenic mice containing large segments of DNA called bacterial artificial chromosomes (BACs), which were inserted into the mice. These large fragments, much bigger than any that had previously been inserted into the mouse genome, had the advantage of containing all the elements needed to make the transgene operate accurately. These BAC mice ushered in a new era of precision mouse models engineered to enable study of specific brain cell types and their roles in development, behavior, and disease.

I met William in the fall of 1999 and was immediately interested in how his technique might be used to study Huntington's. Right away I invited him to the next Venezuela study visit. He later told me that the trip had been life-changing for him, both seeing the distress of the people with the disease and their families and also meeting the impressive HD researchers on this expedition. He said he was "bitten by the HD bug." Although he obtained his medical license in New York, he decided to go to UCLA to set up his own lab and focus on developing BAC transgenic mouse models to study Huntington's and other neurodegenerative disorders. Very soon his lab produced the famous BACHD mouse. This mouse carries the entire human mutant *huntingtin* gene (*mHTT*), which is expressed (activated) in all the same cell types and brain regions as humans with the disease. It also shows small changes in the DNA (single nucleotide polymorphisms or SNPs) that turn out to be useful for developing therapies that lower only the mutant version of the gene (the mutant allele) while not interfering with the typical wild-type allele.[6]

They also showed that the BACHD mice produce the same mutant huntingtin protein as humans. Moreover, this BACHD mouse exhibits the same behavioral and motor problems as people with Huntington's. Their mouse brains also show similar manifestations as do the brains (at autopsy) of human beings. In short, because the BACHD mouse models the disease so accurately, it has also proved amazingly useful for delineating precisely which types of cells in which parts of the brain are most affected by the altered huntingin protein and along what timeline.

♦ ♦ ♦

Enter Ai Yamamoto, a young neurobiologist at Columbia University whose imagination took everything to the next step—and the next. In 2000, as a graduate student in the lab of neuroscientist René Hen, a few floors above my office at the New York State Psychiatric Institute, Ai and a postdoc named José Lucas along with René Hen, figured out how to use a "switch" to turn off the disease that Gill had turned on. The switch was based on the way bacteria turn certain genes on and off. They put both a piece of the *huntingtin* gene and the switch into a mouse. Then, as the mouse started to develop symptoms, they turned the switch off. The mouse got better. Its brain improved and its movements returned to normal. This phenomenal result suggested that, in humans too, an inherited neurodegenerative disease such as Huntington's might be reversed.[7]

Meanwhile David and I and others homed in on another revelation of the gene discovery: an inverse correlation between the size of the CAG repeats and when symptoms began—the higher the number of CAGs, the earlier the symptoms. Nonetheless, for individuals with 40–50 CAGs—which included the majority of those with the disease—this correlation was much looser. People with the identical number of CAGs often developed symptoms at widely differing ages. Why, for instance, did some people with 44 CAGs become symptomatic at twenty-five years of age whereas others with the same 44 repeats remained symptom-free until they were forty?[8]

Most of the Venezuelan families lived in similar environments, although we figured out that environmental factors, such as diet and exposure to toxins, evidently did play a role. But David's statistical analyses suggested that there must also be genetic modifiers—that is, gene variants unrelated to the *huntingtin* gene. If so, then these other genes might also be targets for intervention.

In 2008, David and I with others published a paper on nine random genetic differences, outside the mutant *huntingtin* gene, that we found among the Venezuelan families with Huntington's, some of which seemed to be correlated with late age of onset.[9] These were chance DNA variants scattered through the genome of individuals whose symptoms began much later than predicted. When I dropped in at Ai's lab at Columbia one day soon after our paper came out, I found Ai and her student Leora Fox discussing it. Ai made an offhand remark about how awesome it would be if one of the variants we had identified occurred in the gene (and protein) called ALFY that her lab had been studying. She said it almost jokingly, but somehow I had the feeling that it could be true, and I insisted she pursue this wild hypothesis that she'd thrown out on a whim. As it happened, one of the late-onset variants we had found *did* turn out to be in the ALFY gene. Ai is overly generous with credit, so she credits me with discovering this property of ALFY, but I say it was another collaboration of scientific data and imagination.

The ALFY protein helps clear brain aggregates. Taking advantage of a garbage system in the brain that acts much like Pac-Man, ALFY seems to help find the clumps and move them into a membrane equivalent of a garbage bag. The collected trash is then carried over to a lysosome, in essence

(Left to right) *Julie Porter, me, and Ai Yamamoto. (Courtesy HDF.)*

the stomach of a cell. Like our stomachs, the lysosome, with its high-acid environment, breaks down its contents and then returns the pieces to the cell to be recycled. This Pac-Man-like pathway, called autophagy, is the way the cell helps to maintain itself, recycling things when they're no longer necessary. Every cell in our human bodies has an autophagy system. So do the cells of bacteria and yeast. The cell chooses what to eliminate by using adaptor proteins like tags labeling the garbage. ALFY is the adaptor protein that labels unwanted aggregates, and it seems to label aggegates that form in different diseases, not just Huntington's clumps, but those associated with Parkinson's, ALS, and Alzheimer's as well.

Ai's lab is currently exploring methods of increasing ALFY in human cells—in the lab, not yet in living humans. But if they succeed, ALFY could conceivably become a treatment to delay the onset of all these diseases. Instead of curing Huntington's, maybe we could eradicate it by postponing it to a far-off future. Science fiction with the possibility of becoming science fact.

♦ ♦ ♦

By the late 1990s, I found myself facing new personal challenges. The fact that I lived in New York City while my father and the foundation office at this time were in Los Angeles created delays and communication difficulties. And the high value we placed on creativity and imagination sometimes led us to neglect the importance of efficiency and transparency, to the frustration of colleagues and even to ourselves. I also found it difficult to turn down invitations, a lifelong problem, to the point that I often found myself overcommitted and falling asleep as soon as I was alone in front of my computer or riding in the car with Herb.

My father's authority in the foundation as chair of the board of directors also didn't sit well with some, who voiced their objections in a way that I found offensive. So what if Dad was in his nineties and growing frail? I adored my father and clashed with anyone who wanted to unseat him from what I considered his rightful position, even if I sometimes clashed with him myself.

But as the years went by, even he, my beloved, seemingly indestructible father, who spent hours listening to enormous novels by Trollope and Dickens and remained intellectually sharp, admitted that he needed more help. Macular degeneration was stealing his eyesight, although he could still see enough to navigate his apartment. "I'm going to simplify everything so that I can memorize where things are and find them when I need them," he would say. And he did just that, even as he was diagnosed with COPD and his medical problems mounted and he eventually needed full-time care. He would live to be 98, still taking walks with friends in Palisades Park in Santa Monica. He continued seeing a few patients from time to time and was always interested in the work of the

*Dad and me, ca. 2000. (Courtesy Elaine Attias.)*

foundation and in the latest Huntington's research. Many friends visited. Frank and Berta Gehry came by often, helping to keep his spirits up. And Arthur Golding, a retired vascular surgeon and friend, came to the rescue more times than I can remember. Miriam Wosk, an artist friend who lived close by, never lost patience with Dad's ever-slowing pace, nor did his analyst colleague Gerry Aronson.

Even on his last day, Dad was up and out of bed. And then he was in his bed that last afternoon, talking about how we would all have herring and port, his favorite aperitifs, at 6 p.m. As his breathing slowed, Michael Rawlins, who happened to be in Los Angeles at the time and came by to visit, took the measure of his heart rate with a stethoscope we kept in the bedroom as I tried desperately to call him back. "Daddy, come back! Daddy, don't die!" Alice and I were next to him on his bed as he took his final breaths.

♦ ♦ ♦

My friend Elaine May told a story about Dad that I have thought of often since he died. She was editing her first movie, which was in trouble. So, on a Friday afternoon, the studio asked her to put something together for a screening by Monday and gave her a weekend editor to work with. But the editor, feeling ill, left in the middle of the weekend, although not before warning Elaine not to touch anything while he was gone. She immediately felt paralyzed and unable to work. Yet she knew how to edit, and everything depended on Monday's screening. What to do! She went across town to see my father, who listened as she described her situation.

Dad: "This is an important screening?"

"Yes," Elaine said, "It could decide everything."

"And you won't finish if you don't start putting scenes together now?"

"Right. I shouldn't even have taken the time to be here now," Elaine replied.

Dad told her she was having a panic attack and suffering from hysterical paralysis, which made it impossible to work on the movie. His advice was to do it anyway.

"Do it anyway? That's your advice? I just told you it was impossible."

"I understand," he said. "But if your entire film hangs in the balance, you have no choice."

So, she went back to the studio and started working, despite her trembling hands and feelings of panic. After a while there was a tap on the door. And there was Dad. "Here," he said, handing her a ham sandwich. "I figured you wouldn't have time to go out for lunch, so I brought you something to eat."

She never forgot that ham sandwich or his "Do it anyway" advice, and neither have I. It wasn't Nike's "Just do it." Dad's key word was *anyway*: Do it anyway, in spite of the doubt and the panic and the pain.

Elaine did. And I do.

CHAPTER

# 10

# LIVING WITH HUNTINGTON'S

For a long time, I had the feeling of being watched. Not just looked at but examined, scrutinized, surveilled, and studied in the way that a clinician looks at her patient. Perhaps it began with the gene discovery. It was as if people were searching me for subtle signs that Huntington's symptoms had begun. I remember that over the years many people I barely knew—journalists, colleagues, acquaintances—took it upon themselves to ask me if I had gotten tested and, if so, what the test revealed. Why did they think it acceptable to inquire about such a sensitive matter? I told them my decision was private and one that I did not care to disclose. Some journalists and even some medical professionals took my reply as evasive, as if I were somehow dodging "truth." Over time, though, the discussions of testing grew more nuanced, as the downsides of predictive gene testing, not only for Huntington's but for other incurable cognitive disorders such as Alzheimer's, became more evident and a majority of people at risk for Huntington's chose not to get tested. Alice decided to answer the queries by stating that she was one of the latter. She wanted to validate not getting tested as a legitimate choice, although one that could change at any moment as promising clinical trials sought tested enrollees. But I wanted to make clear that the questions were intrusive. It was none of their business after all.

I did sometimes ask myself to what extent I may have been influenced by my father's opposition to our taking the test. He certainly didn't hide the fact that he was against it, even while acknowledging that it was our decision to make. Our tangle with him just after the marker discovery had impressed upon me the extent to which genetic testing is not just an individual matter and that it can have profound implications for everyone in the family, psychologically at least. I knew this intellectually, from my research and my clinical practice. Now I was experiencing it at the deepest emotional level.

My partner, Herb, was neutral on the subject of the test. Fortunately for me, he didn't press or encourage me one way or the other. He was happy to follow my lead and support me, whatever my decision might be. I preferred to live with uncertainty and ambiguity. I wasn't sure I could live with the foreknowledge that one day the dreaded symptoms would emerge. I preferred to meet the devil when he was on my doorstep. Nonetheless, during the years of our visits to Venezuela, I would ask Anne to give me a neurological assessment, just to be sure I was OK. And each year she would assure me she didn't notice anything amiss.

Alice tells me that in 1996, three years after the gene discovery, she thought she noticed subtle symptoms, little movements, just enough to comment on them in her diary. That was the start of her worrying. I was fifty-one at the time, a little younger than Mom when she ran into that scolding policeman who shamed her for allegedly being drunk as she was on her way to jury duty.

Two years later, my beloved Anne confided that some of our team members had been speculating about my having Huntington's. She told me *she* hadn't noticed anything, however. There evidently had been telephone calls and gossip. I was upset and angry about people talking about me behind my back. After all, these people weren't my doctors. I wasn't their patient. They had no business trying to diagnose me outside of the clinic.

Even some of my other good friends took it upon themselves to tell me the news! Mike Conneally, my pal from the commission and one of the Gene Hunters, whom I loved dearly, said to me point-blank, "Nancy, you have Huntington's." So did Peter Harper, another Gene Hunter and future Sir Peter, who announced to my sister at a Huntington's meeting in the UK that people with HD tend to do better when they accept the diagnosis early on in the disease. I remember the outraged response of Michael Rawlins, also soon to be knighted, who was at that UK meeting and confronted Peter on his out-of-the-clinic diagnosis, dressing him down for interfering with both Alice and me.

Unfortunately, the gossip continued. At workshops I could feel people watching me closely, trying to determine how still or wiggly I was. I knew they sometimes discussed what medication I might be taking that would account for differences during the day. There were times in the early 2000s when I, too, was startled to see myself in a video or a filmed interview, noticing myself moving around more than I had perceived. Anne acknowledged that now many people were asking her about my status. She told them she was my friend, not my neurologist. And there were others, too, who made that point strongly, such as John Mazziotta, a neurologist and pioneer in brain mapping at UCLA, who presided over the Hereditary Disease Foundation's Scientific Advisory Board for many years. John was great at tossing the question back at the questioner, as was Richard Mulligan, from our Gene Hunters advisory committee, and quite a few others. They understood that making diagnoses outside of the clinic was inappropriate and unethical—and damaging to boot. If I was unable to carry out my responsibilities, that was what they should focus on. True, there were moments when I, too, worried about my brain, as I told Alice one November day in 1998, when we took a long walk on the beach in Santa Monica. (I have no recollection of that day, but she noted it in her diary.) Evidently, I told her how I couldn't stop myself from spending too much money, buying too many things, being late on credit card payments, and paying penalties. I knew I was overextended but I couldn't say no. Without our foundation administrators, Judy Lorimer and Julie Porter, I would have been lost.

I decided to see Linda Lewis, a movement disorders neurologist at Columbia, whom I respected and liked. I asked Linda to check me out, although I did not ask her for a diagnosis and she did not give me one. I had had my first hip replacement around that time, and I wanted reassurance that this surgery was the origin of changes in my gait and stance. She told me that it might well be.

Throughout this time, Herb insisted that he didn't notice any symptoms. He was good at denial too. He focused instead on the state of our apartment, which somehow never managed to get itself in order. Things were clean but chaotic, to the point that Herb, helpless to do anything about my clutter over the years, decided finally, just before the COVID pandemic shut down the world, to buy a new apartment. Herb told me to take with me only what I needed and to leave everything else behind. And I did. But still I asked myself, which was more important: a neat and tidy apartment or trying to cure Huntington's? It wasn't even close.

♦ ♦ ♦

While I was spending more and more time focused on the Hereditary Disease Foundation, an advocate in the UK named Charles Sabine began spearheading an initiative to combat HD's stigma. This 2017 endeavor led to one of the most meaningful moments of my life. Charles was a courageous former NBC television war correspondent who had reported from many battlefronts and other dangerous zones. None was as frightening, he used to say, as going through the presymptomatic genetic test for Huntington's and learning that he carried the aberrant gene.

Charles's excellent idea was "Hidden No More," an effort to challenge the continuing secrecy and stigma surrounding the disease. With the help of a brilliant Italian scientist and lifetime senator in the Italian parliament, Elena Cattaneo, Charles organized an audience with Pope Francis in May of 2017 for members of families with Huntington's from all over Latin America and especially from Venezuela.

*With Pope Francis and Elena Cattaneo at the Vatican in 2017. (© Vatican Media, reprinted with permission.)*

I, too, traveled to Rome for this occasion. Standing in the Papal Hall of the Vatican with several hundred HD family members—many from Latin America—I listened to Pope Francis speak about the "tragedy of shame, isolation, and embarrassment" that those with Huntington's experience. He emphasized that "Hidden No More" isn't a slogan so much as a commitment. I especially appreciated his statement that disease "can also be an opportunity for encounter, for sharing, for solidarity." He added that "the sick people who encountered Jesus were restored above all by this awareness. They felt they were listened to, respected, loved." He spoke about the importance of "travel companions" who accompany their loved ones on their difficult path, a path that can be an uphill climb. Like everyone in that audience, I felt that Pope Francis spoke directly to me and connected me with all the others present in that great Papal Hall. And then he asked us to pray for him as he would pray for us. Imagine! The Pope asking *us* to pray for *him*! After that, he came down into the audience and greeted each one of us individually, listening carefully to the words we each spoke. It was as if, for that moment, each of us was the only one present. He even submitted to one of my hugs and gave me one in return.

♦ ♦ ♦

As I write this book, thirty years have passed since we discovered the *huntingtin* gene and forty since we found the marker. I can't help thinking of what has and has not changed in my life since the day of that historic call from Jim Gusella, his voice quiet over the phone saying we had found the gene.

Herb is now gone from this earth, although his presence fills the rooms of our apartment. Against all the predictions and claims of my father that neither Alice nor I would develop Huntington's, that 50–50 probability turned against me. It wasn't supposed to happen to either of us. Perhaps especially not to me, since I had devoted my entire career to trying to find a cure. When I saw myself in video interviews, I noticed the movements. I could tell what was happening. But for a long time, I didn't accept it. Denial was my mode of operation, carrying on the family tradition. I didn't want to be considered "diminished" even as I spoke out against the stigma associated with this disease and disabilities of all kinds. After all, I was still going to meetings, giving talks and interviews, engaging with the foundation, and traveling with Herb and his kids and grandkids on vacations.

But by 2019, the possibility of entering a clinical trial sponsored by the pharmaceutical company Roche served as an impetus to confront a diagnosis. This clinical trial of a drug called tominersen was the first human trial for Huntington's disease of an ASO (antisense oligonucleotide), a molecule designed to interfere with the production of the huntingtin protein and thereby lower the toxicity delivered to the brain. I went to see Linda Lewis again and this time, I asked her directly, do I have Huntington's? And she said, "Yes, Nancy, yes you do." I decided to wait until Herb came home that night to tell him. It was devastating, but not surprising. He held me and we cried together.

My next step was to tell the board of directors of the Hereditary Disease Foundation at a late afternoon meeting on February 19, 2020, at the Milstein Hospital, part of NewYork-Presbyterian up on 168th Street, off Broadway, where Herb had his office. I remember that not everyone could come that day, so the group was small. Alice had flown in from California to give me moral support. We sat around a long table in a small, spare meeting room, just off Herb's big main office. I held his hands, both of them, as I told everyone about my diagnosis, trying not to cry too hard. They were sympathetic, of course, but also not surprised. Afterward Herb, Alice, and I went to The Palm, an old-fashioned steakhouse on 50th Street, for a late dinner. It was a long and difficult night, but we managed to get through it together.

At that point I decided to come out publicly about having Huntington's, hoping that revealing my diagnosis might help others in my shoes. Over the years I had counseled so many people: There is no shame in having this disease. But I had resisted acknowledging it in myself. Now I wanted to make clear that, painful as the illness was to accept, I was still living a full life.

A few days later a science writer I admired at *The New York Times,* Denise Grady, interviewed me. And on March 10, 2020, just before the COVID pandemic shut down the entire world as we knew it, the *Times* published her gracious essay on the front page of the Science section, together with the German photographer Peter Ginter's photo of me holding one of the Venezuelan kids whom I loved, now long gone.[1] I never entered the Roche trial, being beyond the age limit for participants, although Roche was considering the possibility of making an exception for me. Ultimately the drug succeeded in lowering huntingtin, but most of the patients did not improve and some fared worse, a crushing disappointment to the HD community. Roche is now exploring whether changes in dosage and/or testing the drug with specific categories of patients might reveal benefits still unknown.

As I write, new stories emerge: about engineering stem cells to replace our destroyed or damaged neurons; about creating safer, more efficient vehicles to deliver drugs to the brain regions and cells that need them most; about ramping up genetic modifiers to strengthen the brain's resilience or reduce the toxicity of the mutant huntingtin protein; about stabilizing those unstable CAGs and reining in their expansionist moves; and about new forms of packaging drugs into pills or shots and of measuring their ability to help or to harm.

♦ ♦ ♦

An interviewer once asked me if I'd ever thought of myself as heroic. I replied that I thought of myself as being constantly overwhelmed. There are ten thousand things I haven't done, I told him. I thought of the Venezuelan HD families as heroic. But starting with the 1983 marker discovery, my shelves started filling up with little statues, medals, plaques, pendants, ribbons, and other trophies in recognition of our work. I was thrilled, in 1987, to receive an Honorary Declaration from the Community of San Luis, the central focus of our efforts in Venezuela. Two years later, the governor's office of the State of Zulia gave me an award, perhaps through the scheming of the amazing Margot. These Venezuelan recognitions moved me deeply because they felt like a validation of our presence, a sign that we were welcome there. I felt these awards also honored the HD families with whom we worked. It was as if those who had been shunned and stigmatized could hold their heads a little higher.

And then my own country started playing catchup. It was dizzying for a while, with several honorary doctorates and other awards following in close succession. In 1991, the New York Medical College made me an Honorary Doctor in Humane Letters. That same year, the year of the Fragile X discovery, my alma mater the University of Michigan

*Michael De Bakey* (far left) *shown with 1993 Lasker Award winners* (left to right) *Günter Blobel, me, Paul Rogers, and Donald Metcalf. (Courtesy the Lasker Foundation.)*

awarded me an Honorary Doctorate in Science, a huge thrill. The pinnacle came in 1993, the year of the gene discovery. Not only did I receive an endowed chair at Columbia, becoming the Higgins Professor of Neuropsychology, but also I was astonished to receive an Albert Lasker Public Service Award. Hillary Clinton gave a beautiful short speech at the award ceremony, and I got to meet the former First Lady.

Shortly after Dad died in 2007, I was overjoyed to receive the Benjamin Franklin Medal in Life Sciences. Ben Franklin was one of my father's great heroes, a man he admired as much for his psychological insight as for his diplomatic skills. I wish Dad had stayed around long enough for that one.

Twelve years later Cold Spring Harbor Laboratory awarded me the Double Helix Medal, presented during a gala at Mom's old stomping ground in New York City, her beloved American Museum of Natural History, in the Milstein Hall of Ocean Life, with a fabulous model of a gigantic blue whale floating above us. Mom would have loved it.

♦ ♦ ♦

As I sit in my recliner on the forty-second floor of our apartment building on Manhattan's Upper West Side, looking out over the Hudson

*(Courtesy of the Franklin Institute.)*

River, I travel back in memory nearly half a century to San Luis, to Barranquitas, to Laguneta, that fragile shimmering *pueblo de agua* that is no more, vanished from the lagoon where once the tiny houses on stilts dotted the waters a few hundred yards from that wild shore. In my mind's eye, I see all of those who once greeted us on their porches and from their *chalanas* on our annual spring visits. Now long gone. But not forgotten. Never forgotten. The courage of those kids and the commitment of their parents to improving the lives of their descendants have made possible much of the progress of the last forty years in understanding our disease. Although we have not yet found a treatment that will stop Huntington's in its tracks, new approaches have opened up possibilities not even dreamed of forty years ago. We have shifted the narrative. Science is always a work in progress, and we are creating better ways to live *with* Huntington's even as we invent new paths toward a cure.

Pursuit of that cure has given shape and richness to my life, brought me many friends, taken me to far-off places, and given me joy as well as sorrow. I've known the thrill of scientific discovery, that ecstatic moment when decades-long efforts to solve a mystery suddenly yield the solution. I am proud of the collaborative culture we've created within the worldwide Huntington's disease research community and the camaraderie we have encouraged across oceans and generations. I've reveled in fostering new approaches in science, in mentoring young investigators and creating connections where none existed before. And I'm grateful for the opportunities I've had to offer solace and support to families struggling with Huntington's and to ensure that new discoveries will yield benefits and avoid harms.

And so I tell my story as an affirmation of life, my own and that of many others, of an entire community dedicated to breaking the chokehold of this disease. I awaken once more to another day of working with my collaborator Mark, who reads this manuscript aloud while we make corrections and revisions; of welcoming Sean, my physical therapist who entertains me with Ethel Merman and with Stephen Sondheim songs while making me do leg raises and squats; of Zooming with Alice, Anne, and Jenny; and of greeting my aides, RN Rose, Gina, and Jean, who know exactly what I need and bring their skills and knowledge every day.

I write these pages so that those who come after can know what we did and how we did it and what we were thinking and how we felt—how I felt—while we were doing what we did. I write so that you may profit from our knowledge and experience, our thoughts and emotions, so that you may take up the challenge where we left off, ask new questions, design new experiments, create new tools, run new trials; so that those who come after may have different destinies, so that their lives may transcend ours.

*Diane Sawyer: "What is your fantasy about a cure? What's your dream of what you will do on the day you get that cure?"*

*Nancy: "The day there's a real cure, I think first of all I would get a huge jumbo jet, and I'd go down to Venezuela. I think that would be just an extraordinary feeling to see their faces, to say to them, 'This is the magic. Open up, and take this pill and you will never have to worry about this disease.' And I imagine this incredible joyousness and everybody hugging each other and crying for joy."*

# EPILOGUE

"Hey darling!" The text from Ed Wild arrived Tuesday, September 23, 2025. Ed is a neurologist and neuroscientist friend at University College London, long-time researcher on Huntington's, and co-founder of the award-winning HD news site *HDBuzz*. "I can't say why, but can you be with Alice on the phone tomorrow morning at 7:30 am, New York time? If you can, you won't regret it!" I wondered why he was acting so mysteriously. "Are you being knighted?" I asked, only half joking. "No, it's much better than that, but I can't explain." I wasn't to breathe a word of this to anyone but Alice.

The next morning at 7:15, a link popped up on my phone to an imminent Zoom meeting with uniQure, the Amsterdam-based biotech company running a clinical trial for an HD gene therapy drug called AMT-130. I could hardly believe their news! Over the course of three years, in a small group of brave trial participants, the drug appeared to have slowed down progression of the disease by 75%! This impressive result meant that people with Huntington's who received the drug might live four years before reaching a level of impairment in their movements, cognition, and ability to manage daily life that, without the drug, would have occurred in just one year. Here was the first indication that it *was* possible to change the course of this seemingly intractable disease.

And there was more. The great advantage of this gene therapy, if successful, is that it is a "one and done" treatment. While the goal of the uniQure trial is similar to that of the Roche trial—that is, to "lower huntingtin" by interfering with the ability of neurons to produce the huntingtin protein—the means differ. The uniQure trial involves a one-time surgery to deliver AMT-130 directly into the brain, specifically into the striatum, the part of the brain most affected by HD. AMT-130 has two components: a delivery system (or vector) and a gene encoding a small RNA, known as a miRNA. The delivery system is based on a non-disease-causing virus that has been changed to carry and deliver the miRNA gene to cells. The gene works in cells by making a small RNA that blocks the *huntingtin* gene from producing both the typical and the toxic forms of the huntingtin protein. By injecting AMT-130 into the striatum, researchers hoped to reduce enough of the toxic protein to protect the neurons most vulnerable in HD (although in this trial they did not measure the amount lowered). *Notably, uniQure designed the gene component in AMT-130 with technology invented at and licensed from Cold Spring Harbor Laboratory after a decade of fundamental research aimed at better understanding RNA biology and gene regulation.*

These positive results from uniQure are preliminary, with only twelve trial participants who received the effective (high) dose. Delivering the drug requires a ten- to twelve-hour brain surgery, which itself carries significant risks. And the great advantage of this gene therapy—enabling the body's own cells to continue producing the drug—may also be a limitation. If something goes wrong, the damage may be irreversible. In its current form, this drug will also be enormously expensive, seriously limiting access worldwide. But the trial demonstrated for the first time that altering the course of a cruel brain disease CAN BE DONE. And that's a game changer. My father's words echo in my mind, "I'm so glad I lived to celebrate this day."

Now we are one step closer to realizing my dream. If new phases of this trial go well, and international agencies approve the drug, a big challenge for all of us will be to ensure that everyone who needs it has access to it, in its current form and in more accessible forms that are sure to come. I think especially of the Venezuelan families with Huntington's and their forebears living around Lake Maracaibo who contributed so much to making this advance possible. Now it is up to us—scientists, companies, insurers, advocates—to make sure that they share in the benefits from all they shared with us so many years ago.

# ACKNOWLEDGMENTS

First, I'd like to express my deepest gratitude to the families with Huntington's in San Luis, Barranquitas, and Laguneta, in Zulia, Venezuela, who participated in the research that is the foundation of this story. Without them there would be no story. Their deep knowledge of their communities and their courage and determination to make better lives for their children and grandchildren were essential to the science that must one day end the scourge of Huntington's disease. I am also grateful to the scientists and clinicians who participated in this research, many of whom later dedicated their careers to understanding and trying to combat Huntington's and helping those in the midst of it to have less stress and more joy. And I give special thanks to those who took time away from busy days to answer questions and share memories for this book. They include (but are not limited to) Ai Yamamoto, Anne Young, Beverly Davidson, Bob Horvitz, David Housman, Francis Collins, Gillian Bates, Jang-ho Cha, Jim Gusella, Leslie Thompson, William Yang, Judy Lorimer, and Julie Porter. Julie read the manuscript several times and provided invaluable editing, corrections, and research, answering endless questions with generosity and aplomb.

Writing can be a lonely process. But I immediately discovered I had an advantage many writers do not have: I am by nature a collaborator. The watershed moments of my life have all come about in concert with others, so many others that there is no room to include them all. In the case of my memoir, I had two people working closely with me, Mark Hampton, a playwright and screenwriter, and my sister Alice, a historian and biographer. I had the gift of collaborators who would help me tell my story without ever telling my story for me. It became my voice in three-part harmony, with me carrying the melody. Early on Alice realized we also needed a professional editor. It was she who contacted

Rowena Rae, whose scientific knowledge and editing expertise helped shape my stories into a book and was a delight to work with throughout.

Steve Uzzell, a *National Geographic* photographer, generously shared his devastating photos taken during the early years of the Venezuela research. And I thank my dear friend Michael Collins for his arresting photos from our Fulbright fellowship time in Jamaica and for many years after. Jeff Szmulewicz, photographer and videographer at NewYork-Presbyterian Hospital, also provided knockout photos from Venezuela and from an HDF event in New York that also grace this book.

Just when we were looking for a publisher, Peter Tarr took it upon himself to contact Cold Spring Harbor Laboratory Press on our behalf, an act of generosity that changed my life. His introduction to John Inglis, the director of CSHL Press, opened the door to the best possible publication experience any writer could desire. Barbara Acosta coordinated many moving parts with grace and understanding; Kathy Bubbeo's precision copyediting combined with superhuman patience exceeded all expectations; Carol Brown managed to secure obscure permissions with efficiency and speed; while Denise Weiss designed not one, but five gorgeous book covers from which we could choose! And John Inglis oversaw everything with a warm and welcoming spirit that made a stressful process a pleasure.

A special thanks to my wonderful home health aides, Gangadai Balgobin, Jean Palisoc, and especially to my guardian angel Rose—Rosalyn Yusah, RN, to everyone else, but Rose to me. Over the years we have been together, she has seen me across the delicate bridge from a totally independent free spirit to a person who needs help with simple daily tasks. In another successful collaboration, Rose and I have created a life for me where my spirit remains free even if I need help getting my morning coffee. Thanks also to Dr. Gail Berry and to Sean Conroy, PT, DPT, who have kept me mobile, in brain as well as body.

Finally, I wish to thank my late beloved partner Herb Pardes, who supported not only the writing of this memoir but also my life and dreams for forty years. We shared the delicate highwire act of maintaining separate and demanding careers while enjoying a full life together. He gave me both emotional security and the freedom to be creative and I miss him every day.

# NOTES AND REFERENCES

## Prologue

1. Five years earlier, Michael Brown and Joseph Goldstein, two biochemists at UT Southwestern, unraveled the origin of a condition called hypercholesterolemia—extremely high cholesterol in the blood causing early heart attacks and strokes—by studying children who had inherited it from both parents. Their discoveries led to the development of drugs, including statins, to control and prevent atherosclerosis.

## Chapter 1: Family Secrets


1. Huntington G. 1910. Recollections of Huntington's chorea as I saw it at East Hampton, Long Island. *J Nerv Ment Dis* **37:** 255–257.
2. 1960. Nancy Wexler author of winning essay for Americanism Contest. *The Palisadian,* January 29, 1960, p. 7.
3. 1961. Palihi pair wins AFS selection. *Palisadian Post,* November 2, 1961, p. 6.


## Chapter 2: Implosion Therapy

1. Milton Wexler to Nancy S. Wexler, n.d., author's collection.
2. Wexler NS. 1979. Perceptual-motor, cognitive, and emotional characteristics of persons at risk for Huntington's disease. In *Advances in neurology: Vol. 23 Huntington's disease* (ed. Chase TN, Wexler NS, Barbeau A), pp. 257–271. Raven Press, New York.
3. I also concluded that psychological tests should *not* be used alone for diagnostic purposes, because they were too unspecific. They could be used to describe the variety of symptoms and help determine what factors might influence a more benign or malignant course of the disease. N.S. Wexler, note 2.
4. Wexler NS. 1979. Genetic "Russian roulette": the experience of being "at risk" for Huntington's disease. In *Genetic counseling: psychological dimensions* (ed. Kessler S), pp. 199–220. Academic, New York.
5. Romi Greenson to Nancy Wexler, December 7, 1973, Author's collection.

## Chapter 3: The Youngest Commissioner

1. All quotes from commission testimony come from 1978. *Report: Congressional Commission on the Control of Huntington's Disease and Its* Consequences, Vol. 4, pts. 1–6. U.S. Department of Health, Education, and Welfare, Washington, D.C.

2. My paper was titled "The counselor and genetic disease: Huntington's disease as a model." https://files.eric.ed.gov/fulltext/ED123536.pdf
3. The commissioners were Marjorie Guthrie, chair; Milton Wexler, PhD, vice chair; Stanley M. Aronson, MD; Ching Chun Li, PhD; Guy McKhann, MD; Alice E. Pratt; Lee E. Schacht, PhD; Jennifer Jones Simon; and Stanley Stellar, MD. I was Executive Director; Charles MacKay, PhD was Deputy Director; and Elizabeth McDonald was administrative assistant.
4. 1977. *Senate hearings before the committee on Appropriations: Huntington's Disease,* p. 81. U.S. Government Printing Office, Washington, D.C.
5. 1977. *Report: Commission on the control of Huntington's disease and its consequences,* Vol. 2, p. 119. U.S. Dept. of Health Education and Welfare, Washington, D.C.

## Chapter 4: Learning from Laguneta

1. Botstein D, White RL, Skolnick M, Davis RW. 1980. Construction of a genetic linkage map in man using restriction fragment length polymorphisms. *Am J Human Genet* **32:** 314–331.
2. N.S. Wexler, "50–50: genetic roulette," unpubl. ms., author's collection.
3. Kravitz E. 2024. *Lobsters and fruit flies and me, oh my!,* p. 221. Self-published.
4. N.S. Wexler to "Dearest Padroncito," Mar. 1, 1980, author's collection.
5. Ibid.
6. M. Wexler to "Dearest Nan," Feb. 7, 1980, author's collection.
7. M. Wexler to "Dearest Nan," Mar. 24, 1980, author's collection.
8. Americo Negrette, "La Catira [The Blonde]," unpubl. ms, n.d., author's collection.
9. Negrette A. 1962. *Corea de Huntington: estudio de una sola familia investigada a través de varias generaciones [Huntington's chorea: a study of one family through several generations*]. Universidad de Zulia, Maracaibo, Venezuela.

## Chapter 5: The Oracle of DNA

1. The RFLP marker was called G-8 after Ginger Weeks, the technician who had made it, in recognition of her dedication to the task. Ginger worked in Jim Gusella's lab at Harvard Medical School, where his team had been developing new RFLP markers. Each RFLP has a specific home or address on a chromosome, just like a gene does, and Jim and his group tested each new marker they developed to see if it traveled with HD in a family. Ginger designed the eighth marker, which came in four forms: A, B, C, and D.
2. Wexler NS, Conneally PM, Housman D, Gusella JF. 1985. A DNA polymorphism for Huntington's disease marks the future. *Arch Neurol* **42:** 20–24. doi:10.1001/archneur.1985.04060010026009
3. There was one other important limitation to the linkage marker test: It required other family members to be tested in addition to the individual seeking a genetic result. Only an estimated 20% of the 150,000 people in the United States living at risk had enough living relatives for the test to be informative.

4. N.S. Wexler to M. Wexler, July 9, 1985, author's collection.
5. N.S. Wexler to M. Wexler, July 5, 1985, author's collection.
6. Wexler NS. 1992. The Tiresias complex: Huntington's disease as a paradigm of testing for late-onset disorders. *FASEB J* **6:** 2820–2825.
7. Negrette A. 1962. *Corea de Huntington: estudio de una sola familia investigada a través de varias generaciones [Huntington's chorea: a study of one family through several generations]*, pp. 220–221. Universidad de Zulia, Maracaibo, Venezuela; Gonzalez AA, 1983. El Mal de San Vito: Avanzando sin control, *El Zuliano,* March 21, 1983, Maracaibo, Venezuela.
8. Would we make the same decisions today? Since 1981, when we began our study, and even since 2002, when our Venezuelan study visits ceased, there has been much discussion within the worldwide biomedical research community of how to handle individual research results with vulnerable populations in an ethical way. Protocols have changed. Were we to begin a new study or a clinical trial, we would certainly follow these new protocols, aiming to deliver the greatest benefit possible to all participants, as defined *in their own terms,* while avoiding unanticipated harms.
9. N.S. Wexler to S. DeMocker, July 11, 2002, Nancy S. Wexler Papers, Special Collections, Health Sciences Library, Columbia University, College of Physicians and Surgeons.
10. The organizations involved in establishing guidelines were the Huntington's Disease Society of America (HDSA, successor to CCHD), the World Federation of Neurology, and the International Huntington Association. The commission had drawn up general guidelines but these later guidelines were more detailed and specific.
11. See Anderson KE, Eberly S, Marder KS, Oakes D, Kayson E, Young A, Shoulson I, PHAROS Investigators. 2019. The choice not to undergo genetic testing for Huntington's disease: results from the PHAROS study. *Clin Genet* **96:** 28–34.

## CHAPTER 6: WILL THE CIRCLE BE UNBROKEN?

1. Meyerson A, et al. 1936. *Eugenical sterilization: a reorientation of the problem.* American Neurological Association, New York.
2. Chorionic villus sampling is a prenatal test for genetic disorders. The chorionic villi are small projections from the placenta with the same genetic makeup as the fetus.

## CHAPTER 7: THE GENE HUNTERS

1. Initially Charles Cantor, from Columbia, who helped develop pulsed-field gel electrophoresis for very large DNA molecules, joined the collaborative group with neuroscientist Cassandra Smith, also at Columbia, although they later left, as did Keith Fournier, geneticist and early member from the University of Southern California. Pulsed-field gel electrophoresis is a method of separating large DNA molecules. Also referred to as genomic DNA fingerprinting, this method allows scientists to analyze genomic DNA directly.
2. Curiously, in Fragile X the expanded gene did not produce the condition at younger ages in succeeding generations, although the condition did get more severe.

## CHAPTER 8: THE "CROWN JEWEL"

1. Angier N. 1993. 10-year search leads medicine to elusive gene. *New York Times,* March 24, 1993; Maugh II TH. 1993. Gene for Huntington's disease found. *Los Angeles Times,* March 24, 1993.

## CHAPTER 9: MAPS, MICE, AND MODIFIERS

1. Members of the working group were among the most significant thinkers on the social dimensions of the new genetics. They included Jonathan Beckwith, a molecular biologist and cofounder of activist group Science for the People; Patricia King, attorney, specializing in the borderlands between civil rights, bioethics, and genetic medicine; Robert Cook-Deegan, a clinician then at the National Academies of Sciences, Engineering, and Medicine, specializing in public policy related to genetics, biomedical research, and health; Robert H. Murray, a clinician, then president of the Hastings Center on bioethics; Robert F. Murray, a clinician pioneer in pediatric sickle cell disease research; and Victor McKusick, a pioneering geneticist and early proponent of the Human Genome Project.

2. See Cook-Deegan R. 1994. *The gene wars: science, politics, and the human genome.* W.W. Norton, New York.

3. N.S. Wexler, Statement before the United States House of Representatives, Committee on Government Operations, Subcommittee on Government Information, Justice and Agriculture, October 17, 1991.

4. Mangiarini L, Sathasivam K, Seller M, Cozens B, Harper A, Hetherington C, Lawton M, Trottier Y, Lehrach H, Davies SW, Bates GP. 1996. Exon 1 of the *HD* gene with an expanded CAG repeat is sufficient to cause a progressive neurological phenotype in transgenic mice, *Cell* **87:** 493–506.

5. Davies SW, Turmaine M, Cozens BA, DiFiglia M, Sharp AH, Ross CA, Scherzinger E, Wanker EE, Mangiarini L, Bates GP. 1997. Formation of neuronal intranuclear inclusions underlies the neurological dysfunction in mice transgenic for the HD mutation. *Cell* **90:** 537–548; Roizin L, Stellar S, Webster H. 1979. Neuronal nuclear-cytoplasmic changes in Huntington's chorea: electron microscope investigations. In *Advances in neurology: Vol. 23, Huntington's disease* (ed. Chase TN, Wexler NS, Barbeau A), pp. 95–122. Raven Press, New York; DiFiglia M, Sapp E, Chase KO, Davies SW, Bates GP, Vonsattel JP, Aronin N. 1990. Aggregation of huntingtin in neuronal intranuclear inclusions and dystrophic neuroites in brain. *Science* **277:** 1990–1993.

6. Gray M, Shirasaki DI, Cepeda C, André VA, Wilburn B, Lu X-H, Tao J, Yamazaki I, Li S-H, Sun YE, et al. 2008. Full-length human mutant huntingtin with a stable polyglutamine repeat can elicit progressive and selective neuropathogenesis in BACHD mice. *J Neurosci* **28:** 6182–6195.

7. Yamamoto A, Lucas JJ, Hen R. 2000. Reversal of neuropathology and motor dysfunction in a conditional model of Huntington's disease. *Cell* **101:** 57–66.

8. Very high numbers of CAG repeats correlate closely with juvenile onset. However, people worldwide with Huntington's typically have between forty to fifty CAG

repeats. In this range, age of onset for the same CAG number can differ by as much as twenty years.

9. Gayán J, Brocklebank D, Andresen JM, Alkorta-Aranburu G, US-Venezuela Collaborative Research Group, Cader MZ, Roberts SA, Cherny SS, Wexler NS, Cardon LR, Housman DE. 2008. Genomewide linkage scan reveals novel loci modifying age of onset of Huntington's disease in the Venezuelan HD kindreds. *Genet Epidemiol* **32:** 445–453.

## CHAPTER 10: LIVING WITH HUNTINGTON'S

1. Grady D. 2020. Haunted by a gene. *New York Times,* March 10, 2020.

www.ingramcontent.com/pod-product-compliance
Lightning Source LLC
Chambersburg PA
CBHW051220150426
42812CB00075BA/3498/J

* 9 7 8 1 6 2 1 8 2 5 4 5 6 *